CYBER REPUBLIC

CYBER REPUBLIC

REINVENTING DEMOCRACY IN THE AGE OF INTELLIGENT MACHINES

GEORGE ZARKADAKIS

FOREWORD BY DON TAPSCOTT

THE MIT PRESS   CAMBRIDGE, MASSACHUSETTS   LONDON, ENGLAND

This book was set in ITC Stone and Avenir by New Best-set Typesetters Ltd. Printed and bound in the United States of America.

Library of Congress Cataloging-in-Publication Data

Names: Zarkadakēs, Giōrgos, 1964– author ; foreword by Don Tapscott.
Title: Cyber republic : reinventing democracy in the age of intelligent machines / George Zarkadakis.
Description: Cambridge, Massachussets : The MIT Press, [2020] | Includes bibliographical references and index.
Identifiers: LCCN 2020000400 | ISBN 9780262044318 (hardcover)
Subjects: LCSH: Information society—Political aspects. | Artificial intelligence—Political aspects. | Political participation—Technological innovations. | Democracy.
Classification: LCC HM851 .Z36 2020 | DDC 303.48/3—dc23
LC record available at https://lccn.loc.gov/2020000400

10 9 8 7 6 5 4 3 2 1

To my parents, Constantine and Helena

> There are those that look at things the way they are, and ask why? I dream of things that never were, and ask why not?
>
> —Robert F. Kennedy

CONTENTS

FOREWORD: TOWARD A NEW SOCIAL CONTRACT FOR THE DIGITAL ECONOMY

Don Tapscott

Writing the twentieth-anniversary edition of my 1994 book *The Digital Economy* was a sobering experience. The book was very positive about the "promise of the Internet" and, to be sure, the Net has brought about many great innovations.

But the book had a small section about The Dark Side—things that could go wrong. Re-reading it 20 years later I was shocked to see that every danger I hypothesized had materialized.

Our privacy has been undermined. The digital economy has created a system of "digital feudalism," wherein a tiny few have appropriated the largesse of this new era of prosperity. Data, the oil of the twenty-first century, is not owned by those who create it. Rather, it's controlled by an increasingly centralized group of "digital landlords," who collect, aggregate, and profit from the data that collectively constitutes our digital identities.

Exploiting our data has enabled them to achieve unprecedented wealth, while at the same the middle class and prosperity are stalled.

In 1994 I had hoped that the Internet would create new industries and jobs—and it did for a while. But today technology is wiping out entire industries' employment, and the imminent threat of structural employment is fueling unrest. Trucking, one of Canada's largest sources of employment, will likely be automated within a decade. Digitized networks enable outsourcing, offshoring, and the coordination of labor at a global scale. Within the second era of the digital age—one centered on blockchain technologies, machine learning (ML), artificial intelligence (AI), robotics, and the Internet of Things (IoT)—many core functions of knowledge work, many companies and industries, are in jeopardy.

Yes, we see a new wave of entrepreneurism globally, but our regulations were designed for the industrial economy and hamper success.

The increased transparency enabled by the Internet has also revealed deep problems in society. Canada is learning the truth about the horrific history of our indigenous peoples who, in turn, now have tools to speak out and organize collective action. We also understand deeply how climate change threatens civilization on this planet and people, especially young people who will suffer most, and who are now organizing to reindustrialize the country and the world.

I had hoped the Internet would bring us together as societies and improve our democracies. But the opposite has occurred. Algorithms expose us solely to information and perspectives that confirm our biases. The upshot has been more fractured and divisive public discourse, and democratic institutions are eroding before our eyes as trust in politicians is at an all-time low. Populist rhetoric becomes more appealing in these conditions, and many are vulnerable to scapegoating and xenophobia. The upshot is that there is a crisis of legitimacy of liberal democracy.

To borrow from Paddy Chayefsky, people everywhere are "mad as hell and not going to take it anymore." As such they have become vulnerable to populism, xenophobia, and scapegoating minority ethnic groups, races, and religions. Centrist parties are in rapid decline and extremist right-wing parties from Hungary and Poland to France and Germany are on the rise. Perhaps as unthinkable as the success of Donald Trump is the rise Bernie Sanders, an avowed socialist who has come in second to Joe Biden and the Democratic Party establishment in winning a democratic presidential nomination. The unfolding story is one of growing discontent, with the deepening economic crisis and the establishment that created it.

Conversely, the next era of the digital economy could bring an age of prosperity, with new networked models of global problem solving to realize such a dream. For the last 40 years we've seen the rise of mainframes, mini-computers, the PC, the Internet, mobility, the Web, the mobile web, social media, the cloud, and big data.

In this next era, new technologies are already infusing into everything and every business process. In addition to AI, ML, and IoT, we have predictive analytics, additive manufacturing, autonomous drones and vehicles, and precision medicine enabling entirely new types of enterprise. Foundational to these innovations is the technology underlying cryptocurrencies—blockchain.

To meet these new challenges, the time has come for every country to reimagine its social contract—the basic expectations among business, government, and civil society.

When countries evolved from an agrarian economy to an industrial one, we developed a new social contract for the times—public education, a social safety net, securities legislation, and laws about pollution, crime, traffic, and workplace safety—and countless nongovernmental organizations have arisen to help solve problems.[1]

It is time to update these agreements, create new institutions, and renew the expectations and responsibilities that citizens should have about society. I've spent considerable time working on a framework for such a new social contract and have come to some pretty far-reaching conclusions.

We need new models of identity, moving away from the industrial-age system of stamps, seals, and signatures that we depend on to this day. We need to protect the security of personhood and dismantle the system of digital feudalism. Individuals should own and profit from the data they create. We need new laws for the operation of autonomous vehicles, robots, drones, and technology in our bodies.

Our basic expectations of work are shifting, but our systems designed to support workers have not. Gone are the days when workers might expect to do the same job or work in the same field for their whole careers. Students today are preparing for unprecedented lifelong learning, with the knowledge that technology will likely require them to reimagine their role in the workforce.

In the face of new models of work, we must update our educational institutions to prepare for this kind of lifelong learning and establish a universal basic income to support transition periods, providing a foundation for entrepreneurialism and investing in the potential of our populations.

We must adopt new models for citizen engagement in our government. In AI and blockchain we have found one such model, with the possibility of embedding electoral promises into smart contracts. Meanwhile, it enables secure outlets for online voting and other forms of direct democracy using the platforms voters use every day.

Networks enable citizens to participate fully in their own governance, and we can now move to a second era of democracy based on a culture of public deliberation and active citizenship. Mandatory voting encourages active, engaged, and responsible citizens.

It's also time for business leaders to participate responsibly—for their own long-term survival and the health of the economy overall. Even—or especially—in a time of exploding information online, we need scientists, researchers, and a professional Fourth Estate of journalists to seek the truth, examine options, and inform the ongoing public discourse. We each have new responsibilities to inform ourselves in a world where the old ways are failing.

Are these expectations overly ambitious or even utopian? I think not. To quote Victor Hugo, "nothing is more powerful than an idea whose time has come," and the challenges facing our economic and political institutions today warrant such a change.

The challenges of our era demand audacious solutions. Now more than ever the world needs fresh thinking for new digital age.

That's why I am delighted to write this foreword to *Cyber Republic*—an extraordinary book by George Zarkadakis. Democracy is in deep trouble. Legitimacy is the idea that you may disagree with whoever is in power, but you think the system is a good one. But more and more people are challenging democracy itself, including Donald Trump who says voting is "rigged" and the center of democracy in the United States is a "swamp." Youth voting in many countries is at an all-time low, and according the 2020 Edelman trust barometer, trust in governments has never been lower in modern history.[2]

It's time for change. George makes the case effectively with deep research, strong argumentation, and vivid examples. He calls for a rethinking of our systems of democracy and of the social contract itself.

Like me, you may not agree with everything in the book, but you will find it enormously stimulating and helpful.

There is a "demand pull" to reinvent democracy, amid the current crisis and the new requirements for government. As George persuasively points out, there is also a "technology push" coming from the second era of the digital age—the age of intelligent machines.

It' my hope that many will read the book and that it will help us to catalyze a global discussion. Read on, debate, and take action!

Don Tapscott is the author of 16 books about the digital age, most recently Blockchain Revolution, *with his son Alex, with whom he cofounded the Blockchain Research Institute. He is an Adjunct Professor at INSEAD, the Chancellor Emeritus of Trent University, and a member of the Order of Canada.*

INTRODUCTION

Ten years ago, the small publishing house that I had cofounded with one of my best friends in Greece collapsed. After laying off around a dozen employees, shutting down operations, and paying outstanding debts, I found myself jobless, heartbroken, and penniless. It was not our fault that our business had gone bust. The economy was crumbling all around us. The financial crisis that started in the United States had landed in Europe and was spreading across our continent, like a modern plague. Banks refused to lend, supply chains ran out of liquidity and trust, consumer spending evaporated, while public services were being drastically reduced or eliminated. Hospitals had no money to pay for drugs, and their doctors and nurses were working for months without pay. Unemployment shot through the roof, and the suicide rate exploded. Greece was being reduced to ruins. It felt like the end of the world.

Without much to do, I roamed the streets of Athens, like many thousands of my compatriots, taking part in demonstrations, raising our fists against the "troika"[1] that had imposed such terrifying and unfair austerity upon us. Memories from the brutal occupation of Greece by the German army during World War II fueled powerful anti-European sentiment. Many saw foreign-imposed austerity as occupation by other means. Conspiracy theories abounded that the country's assets were being sold off for pitons. Brazen headlines in the German press such as "Sell your islands, you bankrupt Greeks— and Acropolis too!" did not help.[2] But demonstrators quickly turned against the local political class too. Members of Parliament and ministers, afraid of the citizens that they were supposed to serve and represent, would not dare walk the streets alone and without police protection. The people blamed them as complicit in their suffering. After all, it was the politicians who had agreed to the onerous terms of the country's bailout. Almost daily, protestors would try to storm the Parliament while it was in session, only to be pushed back by the police. The acrid stench of tear gas mixed with that of burned tires hovered around the city for months. These were strange and bewildering times for a country that had suffered under a military junta as recently as the early 1970s, and for a people who had fought for the right to vote for a democratic Parliament.

And while representative democracy was hitting the lowest levels of popular trust, a different kind of democracy emerged in the city's streets and squares, one that ancient Athenians would easily recognize. Citizens formed ad hoc assemblies to discuss, deliberate, and vote on a wide range of issues. Bypassing the cash-strapped government—which was unable to help anyone anyway—citizen groups set up marketplaces where agricultural produce could be sold directly from producers to consumers to keep prices low. With the national health system in tatters, volunteers organized street clinics for those who could not afford private health care. Others would establish soup kitchens and food banks for the needy, whose numbers swelled by the day. Out of the rubble that Greece had become, people self-organized and took whatever control they could of their destiny. Similar grassroots movements of mutual support and direct democracy emerged in other countries that were also hit hard by the economic downturn, like Spain. In Barcelona, neighborhoods ran their own citizen assemblies. The anti-austerity movement Podemos, inspired by the traditions of the Spanish Second Republic, saw the crisis as an opportunity to reinvent city government from the ground up.[3] A few years earlier the banking crisis in Iceland had also resulted in citizens taking charge from politicians, forming a popular assembly and voting for a new constitution.[4] For me, it was particularly fascinating to witness how citizens mobilized and self-organized in times of crisis, and how they reinvented democracy in its rawest, most direct form; for I too had participated in a pan-European citizen assembly several years earlier and had known firsthand the power and social dynamics of direct citizen deliberation and action. Maybe there was hope, I thought.

And indeed, something seemed to change. The direct democracy of the streets was finally making an impact on the representative democracy inside Parliament. The conservative government fell, and fresh elections brought to power Syriza, a far-left party that promised to stand up to the country's creditors and renegotiate a less onerous deal. A referendum followed wherein more than 60% of voters backed the government and rejected the terms of the bailout.[5] There was jubilation in the streets. Five years of demonstrations seemed to be paying off. Unfortunately, those aspirations and dreams were quickly quashed under the brutal realism of international politics: it did not matter how people felt or what they wished or voted for. Others determined our fates; decisions were taken in faraway places, behind closed doors, where money mattered more than dreams and no one cared to ask citizens if they agreed. The new, leftist government was forced into a dramatic turnaround by the hated troika, agreeing to even harsher terms that piled more debt on the country for many generations to come. We, the people, were shown to be powerless. And as the crisis deepened, we only got poorer and more desperate.

Disillusioned and without hope, my wife and I decided to throw in the towel, leave Greece, and begin a new life in England. Little did we know that we were leaving one crisis to be met by another. As we were packing suitcases into our car to drive away, hundreds of thousands of refugees from the senseless and bloody wars in Syria, Iraq, and Afghanistan amassed along the Turkish coast, risking—and often losing—their lives in order to cross the Aegean Sea over to Greece for a better life. It was an irony that defied all ironies. It was also the beginning of the metastasizing of the financial crisis into an existential crisis for European democracy. In the years to follow, the issue of immigration would toxify every political debate and give rise to a virulent brand of populism unseen in Europe since the 1930s. The ghosts of the past had risen from their graves to haunt us. Democracy was under attack by a new cadre of politicians who challenged every principle of "liberalism," the core ideology of Western democracy. What was even more puzzling and disturbing to me was that their attack on liberal democracy was made in the name of democracy! Something was wrong. Nothing made sense anymore. The world had gone mad.

The seeds for this book were planted in my mind as I was crossing a turbulent Europe in my car, all the while thinking about how our lives had changed dramatically because of bad political decisions in which we had neither influence nor say, let alone control. Till then, as a citizen and an entrepreneur, I considered democracy as the indifferent backdrop to my personal, social, and economic life. I was content to leave politics and the running of government to politicians and focus on my family, friends, work, business, and hobbies. I did not really trust politicians, the system was obviously far from perfect, and there were injustices that would often make me angry, but democracy, despite its many imperfections, had the advantage of learning from its mistakes, improving, and evolving—it was, as Winston Churchill famously quipped, "the worst form of government except all those other forms that have been tried from time to time."[6] Furthermore, my generation had come of age at a time when liberalism appeared unchallenged, with all other competing ideologies and systems of government confined to the dustbin of history. Liberalism's successes were many and profound. Global markets were made open and free, as were borders. Hundreds of millions of people around the world saw their income rising and their families coming out of poverty. Infant mortality had seen a dramatic fall, and global illiteracy numbers dwindled. In the West, men and women were free to choose whom they loved without the fear of being persecuted for their sexual preferences. Scientific discoveries, new knowledge, and technological innovation created new opportunities for the future and innovative ways to solve big global challenges such as climate change, environmental degradation, food insecurity, and poverty. Until the crisis came and annihilated our livelihoods, many, including me, felt lucky to be living in free and prosperous times, moving onward to an even better tomorrow.

All those dreams, aspirations, and expectations were shuttered during the "lost decade" that followed the global financial crisis of 2007 to 2009 and the Great Recession. Our politics became polarized, voters moved toward the extremes, and the center collapsed. UK citizens voted to extract their country from the European Union in the 2016 referendum. Far-right, nationalist governments rose to power in Italy, Poland, and Hungary, while nationalist populists effectively framed the political debate in France, Holland, Denmark, Austria, and Germany. The European Union, the most ambitious peace project of the liberal order of the twentieth century, was accused of being the work of self-serving elites.[7] Donald Trump won the 2016 US presidential election against all odds and despite desperate attempts by mainstream media to prevent him from doing so. Trump understood that in our digital age politics has been colonized by show business and voters behave like an audience consuming entertainment. By using the power of social media, he bypassed the authority and critique of the fourth estate and demonstrated how the political campaigns of the future will be fought and won—for better or worse.

And then, like tragic irony, the COVID-19 pandemic arrived in the midst of a new US presidential election year, and shortly after the UK had exited the European Union. Despite the lack of adequate data on the prevalence of infections, most governments decided to hedge against a possibly catastrophic impact on their healthcare systems and put their countries on lockdown. In the spring of 2020 nearly half of the world's population were confined to their homes. The global economy ground to a halt. Authoritarians saw in the pandemic an opportunity to extend oppression and surveillance in the name of public health. But for democracies the pandemic posed a terrible trilemma between curtailing deaths, lifting lockdowns to save the economy, and upholding civil liberties. This trilemma would test citizen trust in politicians and experts even further. How democracies resolve this trilemma may indeed determine their future in the post-pandemic world.

While the forces of authoritarianism, nationalism, and antiglobalization find a new ally in the COVID-19 pandemic, a tsunami of technological disruption is compounding the historical, and unprecedented, challenges to democracy, liberalism, and free markets. Artificial intelligence (AI), together with intelligent robotics, sophisticated sensors, communication networks, and big data, is reshaping the global economy and ushering in a new industrial revolution—the Fourth Industrial Revolution.[8] Just as in the First Industrial Revolution in the eighteenth century when machines automated manual work, this new "intelligent machine age" in which machines are automating intellectual work is forcing us to adopt new forms of economic and political organization.[9]

Just imagine a factory where production is fully automated. Robots do the mechanical work, the casting, the molding, the welding, the joining, the logistics; AI systems perform the design, the planning, the strategy, the pricing, the financial management, the marketing, and the sales. Humans are virtually absent, for they are not really needed. A new class of supersmart robots and several million lines of software code have replaced human dexterity and creativity. Labor and knowledge have thus become tradable assets that can be bought, sold, or rented, instantaneously and at scale. In such a scenario the efficiency of capital is maximized, and capitalism is reaching its apogee by producing goods and services of unparalleled quality at the lowest possible price. But here's the rub: as AI-powered automation makes companies more efficient and productive by eliminating the need for human labor,[10] consumers—the ex-workers in those companies who are now unemployed—no longer have the means to purchase those fantastic new goods and services. Automation kills not only the worker but the consumer too. Capitalism's great success thus becomes its ultimate downfall. To demonstrate the point—often referred to as the "automation paradox"—*The Economist* published an apocryphal story, supposedly taking place in the 1960s, in which two titans of the automotive industry, the industrialist Henry Ford II and the leader of the United Automobile Workers Walter Reuther, were touring a new, fully automated manufacturing plant. "Walter, how are you going to get those robots to pay your union dues?," teased the boss of Ford Motor Company. Without skipping a beat, Reuther replied, "Henry, how are you going to get them to buy your cars?"[11]

Nevertheless, the most politically explosive outcome of full automation is not ending up without consumers but the transformation of a free economy into a zero-sum game wherein capital wins and labor loses. Left unchecked, those who own the robots and the AI systems would enjoy a disproportionate share of the wealth generated by the intelligent machines,[12] while the rest of humanity is relegated to a subsistence level of existence. In a "business as usual" scenario the Fourth Industrial Revolution could make today's inequalities seem petty and insignificant and confirm the most dystopian prophecies for the future. Surely, such a future is morally and socially intolerable.

And while automation looks like a threat to free market economies, a different story unfolds in communist China. There, Marx appears strangely vindicated. For Chinese central planners, AI and robots are what true communism needed in order to, finally, occur. As the state has ultimate power over every business in China,[13] and as Chinese citizens are nudged into sharing their personal data through the "social credit system," the most important economic and geopolitical competitor of the West is moving toward digital authoritarianism inspired by Confucian ideas of social order and harmony. China's spectacular rise over the past twenty years has debunked the myth that capitalism and liberal democracy are the two sides of the same coin. Clearly, they are not. It now appears that communism may be a system better suited for managing the socially devastating externalities of a fully automated economy. That's because Chinese citizens, although they may not have a say on who governs them, will, at the very least, be protected by their presumably benevolent, AI-powered, centralized state if and when full automation hits—more so than their counterparts in the West who, unless something changes, will be facing poverty and destitution. Given how the Fourth Industrial Revolution could drive different outcomes for different political systems, is there a future for liberal democracy in the age of intelligent machines? And, if yes, what should that future look like?

These are the questions that this book will aim to answer. For there is also a very optimistic side to this new wave of cognitive automation technologies, one that can be exploited to advance, not reduce, democracy and the welfare of citizens in a free society. Intelligent machines are cognitive multipliers that can massively augment human productivity and creativity. By freeing us from mundane and repetitive tasks, intelligent machines can allow us more time to spend on things we love, and with those whom we love. Algorithms that make sense of big data can catalyze new scientific discoveries in medicine, physics, biology, materials science, and space travel, to name but a few, and can help provide solutions for global challenges. In order to avert the Fourth Industrial

Revolution becoming a winner takes all, we must reimagine cognitive automation so that the wealth it generates can be shared more widely and fairly. In effect, we must find ways to democratize the digital economy. We are already seeing how such democratization could take place at the level of the enterprise. Artificial intelligence in combination with cloud computing, data, and emergent distributed ledger technologies—that are increasingly referred to as "web 3.0" technologies—could transform how businesses are organized, how work is done, and how the fruits of business success are shared and distributed. Such profound transformation in business already erodes the traditional, hierarchical structures and governance of industrial-era corporations and creates new organizational paradigms wherein collaboration is virtually leaderless, peer-to-peer and interconnected, and wherein governance is more participatory and democratic. Capitalism need not become a zero-sum game because of automation. We can use automation technology in a different way and harness free markets, as well as empower human ingenuity, so that every citizen has a meaningful stake in the future.

In the years following the collapse of my small business in Greece, I had the opportunity to join, as an AI engineer and management consultant, the hundreds of thousands of entrepreneurs, businesspeople, and technologists who are currently shaping the Fourth Industrial Revolution. My job has taken me around the world, from participating in casual meetups of breakthrough innovators, to running workshops on future organizations and discussing the future of AI with executive leaders from some of biggest multinationals, to having conversations with politicians, activists, artists, and lay citizens at public events. My conclusion has been that we in the West are moving toward the future at three different speeds. The avant-garde are the technology innovators; they are the geniuses that are achieving the impossible through hard work, passion, and the shear strength of their imagination. They are moving extremely fast, breaking new barriers on a daily basis, creating with the help of capital the valuable companies of tomorrow. At some distance behind them, and at much slower pace, follow the big corporations. Owing to legacy systems and processes, anachronistic cultures and organizational structures, and industrial-era methods of governance, global players across every industry struggle to catch up with so many waves of disruption coming at them at quick succession.[14] Finally, at the slowest possible speed, is everyone else, including small business, lay citizens, politicians, and regulators. Depending on the speed you travel, you have a different perspective and understanding of the AI revolution.

The biggest risk of this three-speed and fragmented approach is that we will fail to reach consensus on the magnitude and the nature of the task and will end up applying old solutions to solve new problems. If we do so, we will fail to exploit the enormous opportunity of the AI economy to completely transform our lives for the better. How we rethink the AI economy in order to enhance democracy, deal with wealth and income inequality, and empower citizens has become even more of a pressing issue given the devastating economic and social impact of the COVID-19 pandemic. In effect, the pandemic has massively accelerated some of the most adverse ramifications of the Fourth Industrial Revolution—specifically, digital transformation of work, massive unemployment, an increased role for the state, and unhindered surveillance of citizen data. Thankfully, the potential of technology to provide much-needed solutions to get us out of this new crisis is enormous. Artificial intelligence is not just another add-on technology to be subsumed into our traditional systems of government and business. Machines that automate the human intellect can change everything. We are at a tipping point in history, and we therefore need a new playbook on how to deal with that change. This new playbook must connect the dots of technology, ethics, political philosophy, exponential innovation, economic fairness, environmental sustainability, and inclusivity and deliver a new synthesis of ideas.

Cyber Republic was born out of my humble ambition to provide a nudge toward a more concerted, cross-disciplinary, and collective effort in putting together this new, and much-needed, playbook for the Fourth Industrial Revolution. I therefore hope that, by sharing my personal experiences and expertise as a technologist, as a management consultant working with companies that are adopting AI, and as someone who has organized and facilitated a citizen assembly at the European level, I will be adding a useful lens to complement the tremendous work already undertaken across academia, and notably among political scientists, on the intersection of politics and technology. Given the herculean size of the collective task, I will focus, more narrowly, in two main areas of inquiry and exploration where I can bring my experiences and expertise to bear the most.

The first area is why we should make—and how to make—our liberal democracies more inclusive. Given my personal experiences in Greece and elsewhere, I will argue that we must adopt mixed models of direct and representative democracy in order for citizens to regain their trust in democratic institutions as well as reengage with politics in a responsible way. My proposals will examine the use of deliberative models of self-governance, such as the spontaneous citizen assemblies in the streets of Athens and Barcelona, and I will discuss how we can scale such an approach and embed it in the liberal system of government. I will argue that one of the two key problems we need to solve, in order for this mixed model to work, is the way in which direct citizen participation is limited by knowledge asymmetries and time constraints. By "knowledge asymmetries" I mean the knowledge gap that always exists between experts and nonexperts. Complex problems require expert knowledge in order to be solved, but in a democracy they also need the approval and consent of the electorate. One of populists' most favorite all-time shooting targets has been the expert. In a recent example, Michael Gove, a British politician who was one of the leaders of the Leave campaign during the Brexit referendum, when asked to name the economists who backed Britain's exit from the European Union, said that "*people in this country have had enough of experts*."[15] I will argue that the tension between experts and nonexperts can be solved to a great extent by leveraging a combination of machine intelligence and citizen assemblies. Given that taking part in politics takes time, and time needs to be compensated when it is scarce, I will also argue that we need to think of the future as a place where everyone has enough time and wealth to participate in politics, and then use the appropriate sets of technology to get us there quickly.

My second area of inquiry is how to transform the digital economy so that wealth is shared more equitably between innovators, investors, and workers. I will argue that, if we want to preserve our democratic liberties and freedoms, we must democratize the AI economy—the successor to the present-day digital economy. For this to happen, we need to solve the wealth asymmetry in a noncoercive way. We must do so not only for ideological reasons but also because, in a globalized economy, national governments already find it difficult to tax the profits of big, global corporations. This problem will become more acute when these national governments find themselves trying to fund national budgets while their tax base at home is dwindling as a result of automation. We therefore need a bottom-up approach to the problem of wealth asymmetry, one that changes the game of wealth creation at the source and does not require a government to act as the wealth and income redistributor. As I will show, automation may not completely eradicate work, but it will probably destroy most jobs and render most of us, as well as the next generations, part-time workers. This transformation in how we work, earn a living, and pay taxes requires a radically new thinking. I will argue that universal basic income—as most people are thinking about it today—is fiscally problematic and practically insufficient to preserve and improve standards of living for citizens when incomes become uncertain and intermittent. And I will propose alternative ways, whereby citizens do not simply survive at near sustenance levels by becoming dependent on an increasingly intrusive state but are given the opportunity to participate actively in the creation of new wealth and be able to have savings. My proposal will look into leveraging distributed ledger technologies in order to create a new generation of digital platforms and markets where participants share ownership in the value they create though their participation, or simply by provisioning use of their data. I will also argue that for this bottom-up approach to reforming capitalism to work we must also rethink and reengineer how private businesses are governed. Future, postindustrial organizations will look very different from the industrial-era companies we are used to today; they will probably be leaderless and owned, as well as governed, by millions.

The nature of politics is such that not everyone will agree with my suggestions or analysis. Whether you are a progressive or a conservative will very much color your judgment of this book. I therefore think it is important to state right from the start where I stand politically and to identify the values that I will defend. I believe that liberty and individual responsibility are the foundations of civilized society; that the state is only the instrument of the citizens it serves; that any action of the state must respect the principles of democratic accountability; that constitutional liberty is based on the principles of separation of powers; that justice requires that in all criminal prosecution the accused shall enjoy the right to a speedy and public trial, and to a fair verdict free from any political influence; that state control of the economy and private monopolies both threaten political liberty; that rights and duties go together and that every citizen has a moral responsibility to others in society; and that a peaceful world can only be built on respect for these principles and on cooperation among democratic societies.[16]

On the basis of this political position, I will bring my proposals together in imagining a future democratic polity that I will call "Cyber Republic." Think of it as an imperfect yet adequate "paper prototype" for a democratic polity in the spirit of design thinking, whose purpose is to aid discussion, debate, and continuous redesign and improvement. The prototype's aim will be to suggest solutions to three existential challenges for liberal democracy: how we repurpose AI as a human-centric technology that works for all, and how we use this technology—in combination with ideas from deliberative democracy and other technologies—to solve the knowledge and wealth asymmetries. I regard those challenges as existential because I believe that, unless we solve them, citizens will abandon the liberal system of government and the core liberal values that inspire it. I will be discussing aspects of a Cyber Republic throughout the book, but here's the general outline: it is an evolved system of liberal government that incorporates direct citizen participation in policy decision-making, while adopting and using human-centric automation technologies to provide equal opportunities for personal and collective growth and economic development. As nations recover from lockdowns and the global economy reboots, it is my hope that *Cyber Republic* will offer a set of useful tools, ideas, approaches, and suggestions for policy makers, researchers, technologists, civil society groups, workers' unions, and governments, as they formulate the new playbook for liberal democracy and free markets in the Fourth Industrial Revolution and the post-pandemic world.

1

DEMOCRACY VERSUS LIBERALISM

What, exactly, do we mean by "democracy" anyway? Surely, it cannot mean that your property can be confiscated by the state without reason, or that you can be arrested without warrant and sent to a prison camp without a trial. And yet, communist countries of the twentieth century called themselves "democratic republics," as indeed the communist "Democratic People's Republic of Korea" calls itself today, appropriating the word "democracy" to mean something completely at odds with our liberal idea of democracy.[1] Closer to home and our times, Hungarian Prime Minister Viktor Orbán has copied pages from Vladimir Putin's playbook and is transforming his country into an authoritarian, one-party political system.[2] By corrupting the independence of the judiciary, using the state apparatus to punish opposition media and organizations with arbitrary tax investigations, and consolidating power with changes in the electoral law and the creation of a friendly media cartel, Orbán proudly declares Hungary as an "illiberal democracy."[3] Similar illiberal democracies can also be found in Poland,[4] Turkey, and Brazil, where democratically elected populist governments are unraveling liberal institutions in the name of nationalism and the "will of the people." But can such a thing exist? Can a democracy be "illiberal"?

Apparently, the answer is yes. As political scientist Yascha Mounk explains in his book *The People vs. Democracy*,[5] we often confuse democracy with liberalism, and this confusion can be a barrier in forging a broad consensus on the nature of the problem that we are grappling with. So let's clarify those terms and how they relate to one another. Democracy is essentially a decision-making process, or a procedure, whereby a group of people take a collective decision based on majority rule. When there is more than one proposal about what to do, the one that gets the greatest number of votes wins. In a democracy everyone has an equal right to a vote, but that's all. It's as simple as that, and no different from two wolves and a sheep voting on what they will have for dinner.

Liberalism, on the other hand, is not a process but a political and moral philosophy that was born out of the European Age of Enlightenment three hundred years ago. Until then European society was stratified on the basis of family bloodline, which meant that a minority enjoyed a higher social status simply because they happened to be born in a family stemming from the landowning nobles of medieval times. Kings were the ultimate lawmakers and rulers by divine right. Liberalism's fundamental, and revolutionary, idea was that everyone had equal rights simply by virtue of being born a human and regardless of family origin. But once you make that intellectual leap, and remove the divine right of kings as lawmakers, as well as the privileges of the aristocracy, you are faced with the following problem: who should decide—and how should they decide—on the laws of a liberal polity where everyone is equal?

This question is far from straightforward. Having equal rights suggests that all citizens should decide collectively on the laws of their polity by using a democratic process. But liberalism stipulates many other rights beyond the right to vote. These rights include, for example, property rights. Given that not everyone is of equal wealth and that the majority are usually less affluent than a wealthy minority, adopting a democratic process in legislation poses an obvious conundrum: how will the property rights of the wealthy minority be protected if the impoverished majority votes to confiscate their wealth? In a direct democracy it is not only the rich who should fear the whims of the majority. What if the majority, for example, decides that same-sex relations should be punished by imprisonment or death? What if the majority votes in favor of expelling "foreigners," or immigrants, or those whose religious beliefs are different from those of the majority? What about *their* rights? The same problem of protecting the rights of minorities applies to almost everything, including the right of free speech and expression. Using a democratic process to make decisions in a liberal world of equal rights sooner or later leads to inequality, oppression of minorities, and a reduction of liberty and freedom.[6]

Democracy, without the checks and balances of a liberal system of government and its institutions, is inherently illiberal. So let us agree that, when we speak about "democracy," what we really mean is "liberal democracy," which is different from direct democracy or other forms of democracy—as, for example, democratic socialism. Inspired by liberal philosophical values, liberal democracy is a system of government that adopts a representative form of democracy in order to defend citizen and human rights against the tyranny of the majority.[7] But, to do so, liberal democracy must keep ordinary citizens as far away as possible from deciding on laws or having a direct say in the running of government. In a passage of his work *The Social Contract* Jean-Jacques Rousseau beautifully summarized the dilemma of liberal, representative democracy: "The people of England regards itself as free; but it is grossly mistaken; it is free only during the election of members of parliament. As soon as they are elected, slavery overtakes it, and it is nothing."[8]

Rousseau pointed to the main problem of liberal democracy, which is that the people are only sovereign at particular points in the election cycle. Before and after those points our legislative and executive institutions are only indirectly accountable to us citizens. Disdain for the sovereignty of the many is arguably the bedrock of liberal constitutions. Many modern American liberal thinkers, such as Harold Lasswell and Walter Lippmann, have posited that the many, or the demos—the "ignorant and meddlesome outsiders," as Lipmann called them[9]—are not fit for government and should not take part in it. For those thinkers, given the *knowledge asymmetries* between citizens of varied educational backgrounds and intellectual abilities, it is best to leave the serious business of government to an enlightened and educated elite. Liberalism requires that the public must be kept in a zombielike trance between elections, with their political power severely restricted, and only allowed to express their preference on a small number of narrow options during election time. In the presidential election of 2016, the Democratic candidate Hillary Clinton was merely echoing that conventional liberalist worldview about citizen power by referring to Trump's supporters as a "basket of deplorables."[10] The Greek phrase "hoi polloi"—which means "the many"—has a negative connotation in English because it is interpreted and understood in the context of liberal thinking as meaning the "rabble" or the "mob." And yet, that's exactly what a democracy is: the rule of the riff-raff, or the "demos." Disregard for the many inexorably leads to "undemocratic liberalism"—to use Mounk's term[11]—where political elites in liberal democracies ignore the popular will or act against it. When this happens, populism returns with a vengeance to exploit the inherent tension between the democratic will of the many and the liberal idea of preserving equal rights by reducing the sovereignty of the people.

LOSING TRUST IN DEMOCRACY

Take, for example, the issue of immigration in the European Union: in the name of the European Convention on Human Rights,[12] liberals in European institutions resist the popular will that is squarely against increasing immigration from outside the Union. "Undemocratic liberalism" thus prioritizes universal human rights that apply to immigrants and citizens alike, while "illiberal democracy" prioritizes citizen rights that include citizens deciding that they do not want immigrants to enter and settle in their sovereign country. President Trump, nationalist Brexiteers in the United Kingdom, Bolsonaro in Brazil, the Italian,[13] Polish, and Hungarian governments, together with alt-right parties in many European countries, use immigration and nationalism to attack core liberal values in the name of the "people." They are successful, at least for now, because many citizens feel that liberal elites, in the name of universal human rights, have excluded them from the material fruits of globalization and ignored their voice. Clinton's "deplorables" are the millions who came up short in the march toward globalization, saw their way of life annihilated, and experienced their values being ignored or ridiculed. Moreover, many citizens feel that liberal elites are colluding with the owners of capital, the bankers, the corporate bosses, all those who support the opening up of borders to foreign workers and corporations, and those who enjoy relative impunity from the law when things go wrong. The many also have it against technocratic experts, whom they consider to have sold out to amoral capitalists.

As a result, many citizens have ceased to trust expert and scientific opinion. Conspiracy theories proliferate over social media, challenging the truth of scientific discoveries and facts and leaving many of the biggest questions of our modern world open to debate ad nauseam. Does our planet overheat because of human economic activity? Do inoculations cause autism? Do foreign workers decrease, or increase, the economic well-being of a country? Is it better for the United Kingdom to be outside the European Union? Is there a "deep state" that seeks to overthrow Trump?[14] It is an enormous historical irony that, at a time of great scientific advances, it is so difficult for facts and truth to be communicated and for them to be accepted by democratic societies with functioning public education systems. This is the deep crisis in liberal democracy we are faced with, and it shows in the numbers. More than half of the people living in liberal democracies are disillusioned with their political system, according to a recent study published by Dalia Research, Alliance of Democracies, and Rasmussen Global.[15] The study found that 54% of citizens in liberal democracies believe that their voice does not have an impact on political decisions, while 64% think their government does not act in their interest. As Nico Jaspers, CEO of Dalia Research and co-author of the study, said:, "Political systems around the world are currently changing with a speed we have not seen in almost 30 years. Right now the biggest risk for liberal democracies *is that the public no longer sees them as democratic.*"[16]

It is ironic that free markets and open borders have triggered such a deep existential crisis in Western liberal democracies. Much treasure, blood, and ink have been spent during the Second World War and the Cold War to defend liberal democracy and free markets—as if they were inextricable Siamese twins. The military annihilation of fascism and Nazism, followed by the implosion of communism, were interpreted as history's endorsement that open markets, representative democracy, free trade, and the rule of law were the only available path toward prosperity. As a result, the West was almost united in the belief that, as free markets expanded around the globe and raised living standards, liberal democracy would become the point where all humanity must ultimately converge.[17]

And yet here we are today, some thirty years after the fall of the Berlin Wall, witnessing liberal democracy's moment of truth. According to Freedom House,[18] 2018 was the twelfth consecutive year of decline in global freedom, without any signs of this trend abating anytime soon. Seventy-one countries suffered net declines in political rights and civil liberties, with only 35 registering gains. States that a decade ago seemed like promising success stories—Turkey and Hungary, for example—are sliding into authoritarian rule. The Pew Research Center found[19] that 17% of Americans think military rule would be a good idea, while 22% favored a "strong leader". Numbers in the United Kingdom were fairly similar. Younger generations, with no memory of the Cold War, are more susceptible to the sirens of nationalism and authoritarianism, particularly when framed around immigration, terrorism, and job insecurity.[20] People feel that capitalism has released forces that liberal democracy cannot effectively grapple with. As a result, all studies point to a huge crisis of confidence among the electorate towards democratic institutions. At the same time, nondemocratic States, such as China, are successfully managing capitalism to raise the living standards of their people. What has gone wrong with free markets? And why are liberal democracies failing to spread the benefits of capitalism more equitably?

THE FAILURE TO SPREAD THE WEALTH

Following the Great Recession and the anemic growth in workers' wages in developed economies over the past decade, we are now witnessing the steepest rise in income and wealth inequality ever recorded, with the globe's 1% richest owning half the world's wealth,[21] and eight men owning more than 3.6 billion of us.[22] What is particularly shocking is that the world's richest people have seen their share of the globe's total wealth increase from 42.5% in 2008, at the height of the financial crisis, to 50.1% in 2017, or US$140 trillion.[23] The World Economic Forum has identified rising income and wealth disparity as the number one risk and biggest driver of international affairs over the next ten years, adding that it is "fraying the social solidarity on which the legitimacy of our economic and political systems rests."[24] Economic growth in liberal democracies seems to have ceased to work for everyone: the rich seem to accelerate their wealth while the rest of us are either stuck in neutral or falling behind. Put another way, those who own capital are becoming vastly better off than those whose main contribution to the economy is through work.

One reason for this is that, in today's globalized money markets, capital can quickly relocate and be invested anywhere in the world, and it does so in mostly "unproductive" investments (for instance, investment in complex financial products), rather than in equipment or infrastructure, or in investments that increase human productivity[25] and, therefore, wages. Low interest rates due to the quantitative easing policies of central banks further encourage this nonproductive allocation of capital. At the same time, the globalization of capital has created a deep interdependence between myriads of interconnected systems: banks, governments, exchanges, investment funds, commodity markets, supply chains. This interdependence reduces resilience, and the resulting supersystem—what we call the "global economy"—has become prone to unpredictable chaotic phenomena.[26] Massive deregulation of labor markets over the past few years has left workers in the developed world exposed and vulnerable to such chaotic phenomena—or "black swans"—that have a devastating impact on the financially weak. The COVID-19 pandemic—arguably a predictable event—exposed the fragility of globalization by triggering a deep recession, disrupting supply chains, and sending unemployment numbers through the roof.[27] Given the instability of the global economy, work in the West has ceased to be a guarantee for a good living, for raising a family, or for giving back to society. Our parents and grandparents lived at a time when earning a decent wage was possible, as it was possible to put aside savings for a rainy day, for their children's education, for retirement, or for investment. This way of life is collapsing. We are now living in the era of job insecurity and uncertainty. US labor data show that 40.4% of workers are contingent or contract workers, with uncertain incomes and little or no job security.[28] In the United Kingdom one in four families have less than US$130 in savings.[29] Similar trends are evident across the rest of the Western world.

As capitalism and free markets seem indifferent to the wishes, or the will, of the demos, the real tension between capitalism and democracy emerges. Citizens, as research has shown, feel that their voices are neither heard nor taken into account. They see their governments as subservient to lobbies and special interest groups that exert enormous influence on public policy and decision-making. Those special interest groups ensure that markets are neither free nor open, but skewed in their favor. Without access to political decision-making, citizens are frustrated by the lack of political transparency, which explains in many ways why fake news and conspiracy theories abound. Lack of participation and, therefore, responsibility by the many in the exercise of politics systematically corrodes trust in democratic institutions.

THE PRINCIPAL-AGENT PROBLEM

Another reason why liberal democracies consistently fail to make capitalism work for the many rests with our representational system of government. The professional political class, to whom citizens have delegated the authority and responsibility of government, have goals that do not always align with the goals of the many—something that is often called the "principal-agent" problem. In politics, citizens are the principals, and our elected representatives are our agents. Representatives are supposed to act on our behalf and do their best to maximize our welfare. But representatives have their own priorities too, the most important of which is to get reelected. Moreover, they are usually immune to the costs and repercussions of their decisions. By not having "skin in the game," our representatives can pursue their own self-serving agendas to the detriment of collective welfare.

The principal-agent problem in politics becomes more acute when there are multiple "principals"—in other words, when our representatives must make a decision that impacts conflicting interests of the various constituencies they supposedly serve. In such cases the constituency that can exert the most influence on the politician will have its way. This is the basis of lobbying, and it leads inexorably to undemocratic outcomes where powerful and influential minorities advance their agenda and interests while weak majorities fall behind and often bear the costs. The principal-agent problem was seen in action during the global financial crisis and its aftermath. Politicians on both sides of the Atlantic prioritized the interests of a powerful and influential minority called "bankers" and shifted the debts that stemmed from their bad bets onto the shoulders of millions of taxpayers, their children, and their children's children. The moral hazard of those actions still reverberates today and fuels the rise of populism and democratic illiberalism. A recent study[30] by Princeton University and Northwestern University has confirmed how truly "undemocratic" is the political system we call "liberal democracy." By shifting through nearly 1,800 US policies enacted between 1981 and 2002 and comparing them to the expressed preferences of average Americans (50th percentile of income), affluent Americans (90th percentile), and large special interests groups, researchers concluded that the United States is dominated by its economic elite. As stated in the research report, "When a majority of citizens disagrees with economic elites and/or economic interests, they generally lose. Moreover, because of the strong status quo bias built into the US political system, even when fairly large majorities of Americans favor political change, the generally do not get it."

The principal-agent problem is also evident when private businesses interact with governments in the context of regulation. There, the misalignment of goals between government representatives, citizens, and business of various sizes results in an uneven field for competition and, in a worst-case scenario, to public administration corruption. As compliance with government regulation reduces profitability, businesses have an incentive to lobby, or bribe, government officials in order to get exemptions. At the same time government officials have an incentive to make regulation more complicated. Corruption does not manifest only in the taking of bribes. "Revolving doors," whereby public sector representatives get lucrative private sector jobs to run corporate government relations departments, provide another way to incentivize increased, and unnecessary, complexity of regulation.

All those instances of principal-agent misalignment lead to suboptimal outcomes for the general welfare and incur economic costs usually called "agent costs." These costs are ultimately paid by us citizens, either by having to pay more for goods and services, or by having innovative new businesses excluded from competition due to high compliance costs, or by getting taxed onerously so that the public sector get even more bloated and more powerful, or—as was the case during the global financial crisis—by having the private debt of failed banks become public debt. In all those different instances, and more, agent costs reduce our economic liberty. This is why big government and onerous regulation should be resisted on the basis of fundamental liberal values. The only way to reduce the principal-agent problem, and defend economic liberty, is to realign the interests of representatives, business, and citizens, so that everyone has skin in the economic game, and the costs, as well as the gains, are shared more equitably among the players. Later in the book I will suggest ways to use technology in order to get—at least some—economic realignment between agents and principals in the postdigital economy of the twenty-first century.

INNOVATION AND GROWTH

Liberal democracy is far from perfect. It is a continuous balancing act between conflicting interests. Although it includes the word "democracy" in its self-description, it is fundamentally undemocratic, and for good reason. Liberal democracy was first and foremost designed to defend economic liberty. This liberty, however, is constantly undermined by the inherent principal-agent problem of political representation. On reflection, it is quite amazing that a political system with so many faults and internal contradictions has managed not only to survive but also to prosper and beat all rivals—at least until now. Perhaps the best explanation for the continuing existence of liberal democracy is that it has used capitalism and free markets to spawn innovation.

A free market economy, where economic liberties and property rights are protected by the law, encourages the smartest to take risks and explore or invent new things since, if those ventures prove successful, the risk-takers will reap handsome material rewards. Risk-taking and the necessary capital to finance risk are the parents of human innovation. The many, who theoretically hold the ultimate veto in a liberal democracy, should therefore be content with an imperfect system of government, as long as they too benefit from the synergy of capital, risk, and new ideas delivering economic growth. The social contract between liberal democracies and the many is therefore founded on the assumption that innovation will deliver economic growth, which will then trickle down and create new jobs, new opportunities, better public services, and an overall increase in living standards for everyone. So one could argue that liberal democracies may currently be under duress, but given enough time, and as innovation keeps creating enough economic growth, citizens will once again see the benefit of allowing the liberal elites to rule. Nevertheless, if the survival of liberal democracy hinges on sustainable economic growth that can absorb the shocks of economic cycles, it behooves us to examine how hopeful one should be about the future of this system of government.

According to the economist Robert Gordon, in his book *The Rise and Fall of American Growth*, the outlook is rather gloomy, as the age of the great inventions ended in 2004 and we are currently surviving mostly on fumes from past innovations. By looking into twentieth-century market data in the United States, Gordon identified two historical periods of high innovation and economic growth. The first was from 1920 to 1970 when output per hour in the United States rose by 3% annually. Gordon defines this output as "total factor productivity" and considers it a measure of innovation. Essentially, total factor productivity is growth in output minus the extra input of labor and capital, that is, the part of growth that can be attributed purely to innovation. During those golden decades, humanity reaped the rewards of industrial innovation in energy with the exploitation of oil, the expansion of the electrical grid to every home, the invention of chemical fertilizers and plastics, clean water and sewage enabling better hygiene, antibiotics saving lives from infections, and air travel. Our grandparents and parents were the first to buy new household machines that made their lives easier, more comfortable, and more productive: washing machines, refrigerators, electric lights, and automobiles. News became electronic and zipped around the globe with the speed of light. On July 20, 1969, people on every continent watched together in amazement, on television, as Neil Armstrong set foot on the Moon.

A second wave of accelerated growth, although less dynamic than the first, took place between 1994 and 2004, when the world enjoyed the benefits of the Internet. But since 2004, according to Gordon, we have entered an innovation trough reflected in the stagnation of wages and nearly flatlined growth. This trough was made worse with the outsourcing of many blue-collar jobs from Western countries to developing countries because of globalization and the savings to companies from labor arbitrage. Although those savings were translated in cheaper imports and low inflation, good jobs for middle- and low-skilled workers have virtually disappeared. Meanwhile, Gordon says, we are still addicted to oil, airplanes keep flying at subsonic speed, and infections are becoming resistant to antibiotics without our having discovered a new arsenal for our defense. And we are still stranded on this planet, the Moon seemingly further from our reach than ever before. We may all have smartphones that enable us to play our favorite games while traveling on crowded public trains, but humanity remains stuck in enjoying benefits of inventions past.

Nevertheless, a new innovation has now arrived that can change everything and reignite the path toward accelerated growth. Artificial intelligence (AI) is a general-purpose technology, similar to steam power or electricity. This means that it can power and enable the development of many other derivative technologies. Because of this multiplier effect, the economic benefits derived from AI are expected to be enormous. According to research by management consulting company Accenture, AI will double global economic growth by 2035.[31] Most other economic analyses agree with respect to the positive economic impact of AI. For example, PwC estimates that by 2030 AI will have added US$15.7 trillion to the global GDP.[32] Their analysis shows that the impact will be driven by productivity gains both on the supply side, as businesses automate processes and "augment" their workforce, and on the demand side, as AI makes available personalized or higher-quality products and services. So the big political question regarding AI is how much of that expected economic bounty would find its way into improving the lives of ordinary citizens in liberal democracies. Could AI be the groundbreaking innovation that halts the downward spiral toward illiberalism? And if yes, what needs to change in our system of government and our capitalist economy for this to happen? To answer these questions, let us first take a quick tour of what AI is and what it can do now and in the future.

2

MACHINES THAT THINK

The desire to create inanimate objects that behave in an intelligent way arguably stretches back to the beginnings of human evolution.[1] Artificial intelligence is a modern manifestation of this innate human desire to create intelligently behaving artifacts, by making use of the computer. Artificial intelligence's goal is therefore to make computers solve problems we generally associate with human cognition and perception: for example, diagnosing a disease, or navigating through an unknown terrain, or understanding human language, or recognizing a face in a crowd. Broadly speaking, there have been two schools of thought on how to emulate human intelligence in a computer, depending on how each school viewed how the brain acquires knowledge about the world.

For the "symbolic" school of AI, knowledge is the result of logic and is therefore something that emerges by combining a description of the world (the "what," or "declarative knowledge") and a description of how to make inferences about the world (the "how," or "prescriptive knowledge"). Perhaps the most successful systems that came out of this school are "expert systems." A typical expert system for medical diagnosis would have a knowledge base with descriptions of symptoms and of rules that encapsulate how doctors reason on the basis of symptoms and other information to arrive at a diagnosis. Symbolic approaches to AI—the favorite school of AI until the early 1990s—hit a philosophical dead end called *Polanyi's paradox*. Named after the philosopher Michael Polanyi, the paradox makes the observation that much of human knowledge is tacit, that is, it escapes our consciousness and is therefore impossible to explicitly articulate, record, and code in a machine. We know things without really knowing how we know them. This paradox deals a serious blow to coding prescriptive knowledge in any symbolic form. To describe the incredibly complex real world by hard coding it is therefore impossible. For example, a driverless car programmed using symbolic AI would soon run into Polanyi's paradox given the seemingly infinite possibilities and unpredictable situations one may encounter on the road. This realization undermined confidence in the symbolic approach and was the reason why AI went through its so-called winters with funding, as well as interest, drying up. Symbolic approaches in AI are still valuable in specific use cases but have largely relinquished the limelight to the second school of thought that proposes an alternative "nonsymbolic," or "connectionist," approach.

Connectionism follows a biological approach to knowledge and tries to emulate the way the human brain functions at the level of neurons. This approach assumes that knowledge is something that has to be acquired by the machine itself and not be handcrafted, or hard coded, by a human programmer. Therefore, intelligent machines should learn by imitating the functioning of the brain. In 1957 Frank Rosenblatt put forward the idea of *artificial neurons* and built the first "perceptron," using motors, dials, and light detectors. He successfully trained this simulation of a neuron to tell the difference between basic shapes. But his approach was severely limited in recognizing complex patterns given the difficulty, at that time, in connecting many perceptrons together to build a sizable network. Connectionism's time had to wait fifty years for three important developments to take place.

The first development came when Geoffrey Hinton and other researchers at the University of Toronto found a way for software neurons to teach themselves by layering their training. The first layer learns how to distinguish basic features, while successive layers identify more complex features, and so on, in a technique we now call "deep learning." Deep learning is one of the approaches within a much larger framework of techniques and algorithms to make computers learn about the world without explicitly programming them; this larger framework is called "machine learning." The second development in connectionist AI owes a lot to the massive volumes of data that are becoming available as billions of people interact with digital systems and devices and to the belief that one can mine valuable knowledge from that big data. Machine learning is the technology that can deliver powerful new solutions for making sense of massive and varied data sets. Data are absolutely necessary for training deep neural networks, and a lot of data makes for better learning. For example, in 2017 Google made publicly available a data set for training voice recognition systems that would distinguish between different accents. For engineers to train a typical deep neural network to distinguish merely 60 words, more than 30,000 audio clips are needed. The third breakthrough took place when a team at Stanford led by Andrew Ng realized that graphics processing unit chips—called GPUs—originally invented for image processing in video games, could be repurposed for training deep learning systems. To accelerate the rendering of graphics, GPUs perform parallel processing, that is, they split processing on several processing units and recombine the result. By exploiting the parallelism of GPUs, Ng showed that deep learning networks could learn in a day what used to take several weeks.

Nowadays, there are neural networks made up of millions of software neurons with billions of interconnections running on thousands of GPUs. Generally speaking, these networks learn in three different ways. The most common way is called "supervised learning," whereby training data sets are "labeled" by humans; that is, the machine learns by been "told" what its output should be for a given input. For example, you can train a machine to recognize cats by showing it thousands of pictures of various animals, but each time a cat appears letting the machine know that this is a cat. Machines also learn through "unsupervised" training, whereby there is no a priori knowledge on how various data are correlated. In this case the machine discovers correlations in the data all by itself, usually by clustering together data that appear to have a high frequency of common features. Third, there is the method of "reinforcement" learning that was impressively used by DeepMind, a Google company that is one of the global leaders in AI research, to build AlphaGo and AlphaGo Zero. In this case the machine is given a goal (for instance, "maximize score in a game") and is allowed to figure out a strategy for doing so (and thus win a contest).

Hardware, as well as software, is used to improve the accuracy and efficacy of AI systems and to increase the range of their application. Google, for example, has developed special-purpose chips called Tensor Processing Units that accelerate the training and deployment of AI applications that run on their open-source AI framework called TensorFlow. Much of AI processing is nowadays increasingly pushed on the "edge," namely, in the devices that collect data in the field; think of sensors in factories or cities, smartphones and wearables, and just about anything that will be connected in the "Internet of Things." This approach is called "federated learning" and has several advantages. First, it respects user privacy since the machine learning crunches personal data on the device, and no user data need to travel to a distant central server. Second, it provides more personalization and lowers latency; the AI on the edge responds much faster even when the device is not connected on the Internet. Federated learning is expected to become the norm for most consumer-oriented applications as more powerful AI-specific chips become available and are embedded in smartphones and remote devices.

THE QUEST FOR ARTIFICIAL GENERAL INTELLIGENCE

Impressive as the developments in connectionist AI may be, we are still at the beginning. Current AI systems are "narrow" in the sense that they can be trained to solve only for specific problems and domains. For example, AlphaGo can only win in a game of Go but is completely useless at everything else. Moreover, the current approaches in deep learning essentially emulate the pattern recognition intelligence of the brain. Human intelligence is much broader than that. We possess memory and common sense, which we use in order to make sense and take action in a continuously evolving and uncertain environment. We also have feelings and emotions that drive our actions and our interpersonal relations. Current AI systems are nowhere near that level. Nevertheless, the quest for intelligent systems that have capabilities comparable to, or even surpassing, those of human brains continues and, indeed, has intensified.

In the summer of 2019 Microsoft announced an investment of US$1 billion into OpenAI, a research group that was originally founded by Elon Musk and Peter Thiel out of concern that AI may be a threat to humanity. The goal of the investment is to test the hypothesis that a neural network close to the size of a human brain running on Microsoft's Azure cloud infrastructure can be trained to be an artificial general intelligence (AGI).[2] Google, through its acquisition of DeepMind, is also pursuing human-level AI. Both companies approach the problem of AGI as one that requires orders of magnitude more computing power and hope that by brute computing force their systems will be able to crunch through such a wide variety of data sets that they will be able to have general application across many fields.

An alternative approach to AGI looks into advancing biologically inspired hardware, such as "neuromorphic" chips that behave like interconnected groups of neurons.[3] Those chips allow for an alternative approach to AGI that is not based on mathematically manipulating weight functions in neural network algorithms but is based on electrical spikes similar to the actual neural dynamics in the human brain. The most developed experiment in using neuromorphic computing to emulate intelligence is currently taking place at Manchester University in the United Kingdom. Researchers at the SpiNNaker project[4] have developed a massively powerful neuromorphic computer that simulates a billion interconnected neurons. Their ambition is to test their machine in applications ranging from robotics to image recognition and gradually scale it toward interconnecting a hundred billion neurons, which is the number inside a typical human brain. Hybrid approaches, where neuromorphic chips are also capable of training neural networks, are being pursued as well.[5]

And while researchers are trying to crack AGI using our existing understanding of brain function, much attention is also directed toward learning from new developments and breakthroughs in neuroscience. Perhaps one of the most interesting ideas from this field is the "free energy principle" developed by British neuroscientist and brain imaging expert Karl Friston.[6] One of the key observations behind Friston's thinking is that brains consume much less power, and transmit less heat, than electronic computers when doing similar tasks. In other words, biological brains have a way to reduce entropy and, just like the rest of life, they self-organize in the most energy-efficient way.

Inspired by this realization, Friston suggested that the reduction in entropy occurs because living things have a universal mechanism that constantly reduces the gap between their expectations and information coming through their sensory inputs. That gap is what Friston calls "free energy." In effect, Friston is telling us that intelligence is the minimization of surprise! There is a mathematical way to describe this mechanism that is based on a construct called a "Markov blanket." A Markov blanket is essentially a mathematical "shield" that separates a set of variables from other sets in a hierarchical way. So Friston suggested that the universe is made up of Markov blankets inside other Markov blankets, like an endless series of Russian babushkas.[7] His ideas are becoming increasingly influential in the machine-learning community as they provide a general theory of intelligence that is mathematical and testable in a machine. Moreover, Friston's theory allows not only for knowledge processing but for action too. Here's how the theory explains why we act: if, say, my expectation is that I should be scratching my nose and my sensory input tells me that my hand is doing something else, I will minimize the free energy gap by moving my hand and scratching my nose. By explaining action, Friston provides a new way of thinking about the design of autonomous robots and is getting us closer to achieving AGI.

AI AND THE FOURTH INDUSTRIAL REVOLUTION

Artificial intelligence does not need to reach the AGI level in order to have a profound impact on human civilization. The current state of narrow deep learning is perfectly capable of performing complex tasks of *perception*—for example, image and voice recognition—as well as *cognition*.[8] Having machines capable of perception and cognition opens up boundless opportunities for innovation.[9] More importantly, AI can take over and automate many human tasks that require perception and, to a certain extent, cognition. Because of that, AI is already impacting the workplace, and—most significantly—white-collar jobs.[10] As an example, AI trading algorithms have almost obliterated Goldman Sachs's Equity

Trading desk, which was once run by 500 people and is now run by just three.[11] Meanwhile, Goldman Sachs now employs 9,000 engineers and data scientists and is investing heavily in machine learning.

The example of Goldman Sachs illustrates the two opposing effects of technological disruption in the economy. Some workers get "displaced" by the new technology and lose their jobs to the new automation—as the trade desk workers did. But new work is also created that has more high-value tasks and is better compensated. As those higher-value workers—the data scientists and engineers in the Goldman Sachs example—are more affluent, their increased disposable income goes back into the economy to "compensate" those who lost their jobs by creating demand for new services.

This interplay between the "displacement" and "compensation" effect has been observed in past industrial revolutions as well. Take, for example, the invention and popularization of the automobile in the early twentieth century. Till then, many people used horses for land transportation, and there was a plethora of jobs that supported the owning, renting, maintenance, stabling, and use of horses. All those jobs were lost when people shifted to owning and using cars. Nevertheless, the automation of horses created new needs and new opportunities for work, in car manufacturing, as well as in servicing, driving, parking, washing, and selling cars. Moreover, the car created completely new opportunities, for example, for extending cities into suburbs and building new homes and infrastructure, with a resulting multiplier compensation effect rippling throughout the economy of the twentieth century. By "automating" the horse, the car "augmented" most humans to become more productive.

As we look into how AI will impact different categories of workers during the Fourth Industrial Revolution, the debate centers on the dipole of "automation" versus "augmentation" and how the displacement and compensation effects will play out. There will certainly be winners and losers. Take, for example, Google Translate, the online AI-powered translation service offered by Google for free. The system is automating the work of human translators, at zero cost, making them virtually obsolete.[12] At the same time, it helps immigrants and refugees arriving in new countries to quickly learn the language and integrate faster. By "automating" human translators, Google Translate "augments" human immigrants. But how balanced is this equation? In the Fourth Industrial Revolution will human augmentation deliver enough "compensation effect" to overcome the "displacement effect"?

Although much of current economic research is generally upbeat about AI delivering handsome dividends to the workforce by increasing human productivity,[13] it fails to factor further advances of AI technologies, including the possibility of achieving AGI in the next decade or so. As many high-level human skills become automated because of advances in AI and robotics, companies will have a financial incentive to shed even more jobs in order to reduce production costs and increase operational efficiencies. By simple extrapolation, if AI advances linearly—let alone exponentially—it will become increasingly harder to find a decently paying full-time job[14] in the next ten to fifteen years. The greatest innovation of capitalism to date—the automation of the human intellect via AI—is replacing human workers with intelligent machines. With work being the fundamental bedrock of any modern economy, can liberal democracies survive in a world without work?

3

A WORLD WITHOUT WORK

Humans have been out of work for millennia, and they have lived happy and healthy lives despite not having to earn a living. At least that's what archaeology and anthropology tell us. Our Paleolithic ancestors would have been shocked to witness how stressful, and often vacuous, our modern lives have become because of the necessity to work. For we may take work for granted today, even extol it as a virtue, but the truth is that "work" is the most unnatural thing we humans ever did.[1]

The invention of work was a watershed in the political organization of human society and is strongly linked with the beginnings of social inequality. Timothy Kohler and his team at Washington University collected measurements of homes on 63 archaeological sites around the world spanning a period from 9000 BC to AD 1500, a period coinciding with the agricultural revolution, during which free-roaming, nonworking human nomads gradually settled to become earth-toiling farmers.[2] Comparing the different sizes of homes for each settlement and period, they concluded that, as people started growing crops, settling down, and building cities, the rich started getting richer. Interestingly, they also noticed a difference in wealth disparity between the Old World and America. For instance, the civilization of Teotihuacan, which flourished in present-day Mexico around a thousand years ago, was based on large-scale agriculture and yet wealth disparity was relatively low. Geographical happenstance may be the most likely explanation for this difference.[3] Eurasia happened to have more animal species that could be domesticated—such as the ox, for example—while oxlike animals did not exist in pre-Columbian America. The domestication of animals acted as "production multipliers" in the creation of wealth. Owning an ox meant you could plough more land, increase your harvest, and create a surplus. By creating a surplus, you could become a creditor, that is, a "protocapitalist," to someone who did not own an ox and thus kick-start the economic cycle of credit and debt.[4]

ECONOMIC ABUNDANCE WITHOUT WORK

The Aagricultural Rrevolution may have spelled the end of leisure and the beginning of social inequality, but the memory of our simpler Palaeolithic lives was not forgotten. Hesiod, in his poem *Work and Days*, describes an epoch of economic abundance wheren no one had to work, and "everyone dwelt in ease and peace" to a very old age. That was the "Golden Age of Man," when humans lived like gods. The Bible recounts a similar story: the first humans, Adam and Eve, lived without worrying about food or shelter, or about the future. Everything was provided for them. Living in the moment, and for the moment, was all that mattered. But once exiled from Eden their lives became a constant struggle against uncertainty. The Almighty decreed besetting work upon us as a dubious blessing: "In the sweat of your face you shall eat bread till you return to the ground . . ."[5] And although these words from the Bible have been interpreted by theologians as sanctifying work, the sore reality is that—unless you are rich enough to recreate Eden on Earth—you are doomed to work for most of your productive life in order to earn a living, support a family, and save for retirement. Was it worth it? The social inequality, the need to work and constant worry about the future, the traumatic divorce from nature?

For many years the answer to those questions was unequivocally affirmative. Hunter-gatherer life, once you removed the tinted glasses of romanticizing it, seemed far from ideal, a constant battle against starvation and brutal, untimely death. It was therefore assumed that, as soon as people realized that they could fill their stomachs more predictably by planting and domesticating, rather than by risking life and limb chasing after wild prey, there was no way back. New archaeological data, however, are increasingly revealing a different story:[6] the adoption of agriculture, and work, was detrimental to human health and quality of life. Living next to domesticated animals allowed viruses to mutate and infect humans. By taking up farming, our diets became limited and we became less healthy. The lives of hunters-turned-farmers were shortened. Anthropological research into hunter-gatherer societies that persist to this day seems to confirm that. Before transitioning to farming, we were healthier, stronger, and happier, and—importantly—we expended a lot less energy to sustain our lives.

Canadian anthropologist Richard B. Lee conducted a series of simple economic input-output analyses of the Ju/'hoansi hunter-gatherers of Namibia and found that they make a good living from hunting and gathering on the basis of only fifteen hours' effort per week.[7] Moreover, the livelihood of the Ju/'hoansi is very resilient. They have access to and use 125 different edible plant species, each of which has a slight different seasonal cycle, is resistant to a variety of weather conditions, and occupies a specific environmental niche. Compare that food resilience with ours today, dependent as we are on just three crops—rice, wheat, and maize—for more than half of the plant-derived calories we consume worldwide.[8] On the strength of this evidence and in comparison with modern industrialized societies, anthropologists consider hunter-gatherers as "the original affluent society."[9] Why? Because the transition from hunting and gathering to agriculture was also a transition from abundance to scarcity; from a world of plenty where one could just walk into a forest and pluck something to eat to a world where, unless you can pay for your food, you will go hungry. Economic theory tells us that the reason for scarcity is that resources are finite and limited, and therefore some decision must be made on how to distribute and allocate them. Before the agricultural revolution such resource-allocation decisions were made easily among the members of a small and mostly egalitarian community of people. But once we became numerous and started living in cities, decisions about the allocation of resources became more complex. Scarcity is therefore not only the raison d'être for economics but the mother of politics too. For who governs how wealth is distributed in a society determines the political system of that society and, ultimately, its destiny.

WORK AS THE MEANS TO PARTICIPATE IN ECONOMIC GROWTH

Our views on the ethical importance of work have changed over the centuries.[10] The ancient Greeks—as Hesiod's nostalgic reference to the Golden Age testifies—viewed work as a burden and as an obstruction to the contemplative life of a truly free person; indeed, for them, "freedom" essentially meant freedom from work.[11] The opposite of freedom was slavery, where work was coerced and turned humans into something "less" than complete persons. On the contrary, Christianity regarded work as the road to redemption and sanctified it. The Christian faithful are called the "slaves of God." But it was the labor and women's movements that sprung from the social upheavals of the First Industrial Revolution that redefined the *meaning* of work, in the way that most of us understand it today: neither a burden nor a blessing, but as the path toward self-actualization and personal independence. Work has been deeply embedded into the functioning of industrialized societies ever since. It keeps the wheels of our economies turning by transforming working citizens and their families into consumers. It defines the ever-tense relationship between the owners of capital and the providers of labor. More significantly, work is the means by which we participate in the growth in our country's economy.

That is somewhat paradoxical, because we may take economic growth as a given and link it to work, but in truth "growth" is something relatively new in human history. For thousands of years following the agricultural revolution there was hardly *any* economic growth. The vast majority of people worked hard but continued to live in poverty nevertheless. There was simply not enough wealth to go around more broadly. Every time an innovation created an economic surplus, the resulting increase in population quickly consumed that surplus and things went back to the way they were before—in what is called a "Malthusian trap." But something profound happened as the First Industrial Revolution kicked in in the 1800s, a phenomenon called the "Great Divergence" (see figure 3.1).[12]

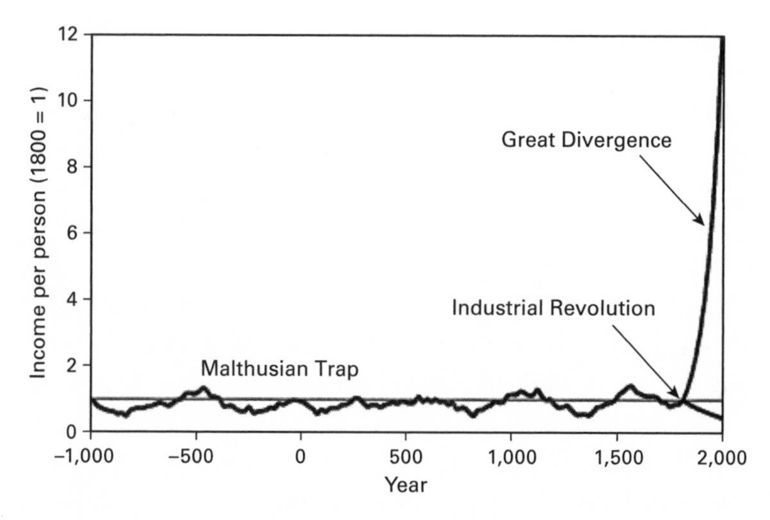

Figure 3.1 The First Industrial Revolution accelerated global economic growth exponentially. Graph based on data from the Maddison project database, by Jutta Bolt, Robert Inklaar, Herman de Jong, and Jan Luiten van Zanden.

Over a relatively short period per capita income increased exponentially. Economic growth—as we presently know it—kicked in. We came out of the Malthusian trap. The First Industrial Revolution created more surplus that people could consume. As a result, the world population started to increase. There were many other drivers for change, ranging from the transition to representational democracy in the West, to the opening up of educational opportunities to the poor, to increased technological innovation. Transportation technologies such as steamships and trains had a decisive impact in boosting international trade and increasing human productivity further. As the latter increased, so did wages and per capita income. To work became the single most decisive factor between staying poor or moving up the social and economic ladder. We may take it for granted nowadays when we think that the economic future of our children and grandchildren is likely to be better that ours, as ours was better than our parents' and grandparents', but this expectation is based on the assumption that human productivity will continue to increase thanks to new technological breakthroughs and innovations, and that working citizens will keep on sharing ever-bigger pieces of an ever-growing economic pie.

Alas, the expectation of continuing economic growth that is shared by the working many is severely tested by data. At least until the 1980s, a stable labor income share was accepted as a fact of economic growth. Over the past decades, however, there is evidence of a downward trend for labor share. According to the Organisation for Economic Co-operation and Development (OECD), the average adjusted labor share in G20 countries went down by 0.3% per year between 1980 and the late 2000s.[13] Work is not as important in inflating the economic pie anymore. Capital seems to play an increasingly more important role. To explore the factors driving this trend, and what could be the repercussions for society, the economy, and politics in the twenty-first century, we must turn to the firm as the main organizer of work, examine how the organization of business processes is changing rapidly because of advances in automation technology, and understand how the role of human work is diminishing.

THE FIRM AS ORGANIZER OF WORK

Nobel Laureate economist Ronald Coase (1910–2013) has offered an economic explanation of why individuals choose to organize by founding a firm rather than by trading bilaterally through contracts in a market. He proposed that firms exist because they are good at minimizing three important costs.[14] The first cost saving comes from *resourcing*: it's less expensive to find and recruit workers with the right skills and knowledge from inside a company rather than searching for them outside the company every time you need them for something. The second cost reduction is in *transacting*, or in managing processes and resources: it's less of an administrative burden to have teams in-house than to manage multiple external contractors. And finally there's a reduction in the cost of *contracting*: every time work takes place inside a company, the rules and conditions are implied in the employment contract, and management does not need to continuously negotiate with individual contractors. By reducing these three costs, Coase suggested that firms are the optimal organizational structures for competitive economic activity.

And yet something seems to be going awry with firms and corporations over recent years. The forces that render firms obsolete appear to have accelerated dramatically. Research by Richard Foster of Yale School of Management[15] has shown that the average life span of firms has been consistently shrinking, from sixty-seven years in the 1920s to fifteen years today, and dropping fast. At current churn rate, 75% of the S&P 500 will be replaced by 2027. We are witnessing a phenomenon of accelerated "creative destruction," to use the terms of Austrian-American economist Joseph Schumpeter (1883–1950). This phenomenon picked up the pace at the dawn of the digital era—around fifteen years ago—by claiming the media and entertainment industries as its first victims. Those industries saw their traditional, analogue business models become obsolete as content became digital and virtually free. What happened to those industries is now taking place in banking and insurance, transportation and retail, health care and consulting, to name but a few. In fact, there does not seem to be a single industry that is immune to disruption by digital technologies. If we focus on how work changes because of digital disruption, then we may clearly see how software is now allowing for the three costs that Coase identified to be minimized *outside* rather than *inside* a firm.

Resourcing freelance contractors with the right skills via online talent platforms—such as Upwork or Topcoder—can be less costly, faster, and less risky than recruiting full-time workers. *Transacting* and managing work is also becoming less costly and more efficient using cloud-based software collaboration tools that enable agile, manager-free forms of work of remote teams, many of whom, if not all of whom, may be freelancers. The advent of blockchain technologies—which will be examined later in detail—promises to slash *contracting* costs by automating contracts with external talent. All these digital innovations are making work easier to resource, transact, manage, and contract out in the market rather than inside a firm. A new form of organization for work, one that is "marketplace-like," is thus emerging: it is not hierarchical but flat, not closed but open, not impermeable but permeable, not based around jobs but around tasks and skills, not run by human managers but facilitated and optimized by computer algorithms, not siloed in functions or lines of business but highly networked and fluid. This new way of organizing work is called a "platform."

PLATFORM ECONOMICS

Let's first define what a platform is and make the distinction between a *computer platform* and a *business platform,* so we may then trace the connection between those two concepts. A computer platform is the environment in which software applications are executed. For example, a web browser is a web-based computer platform where other applications can run: a website, or some executable code, like a document or a spreadsheet. A business platform is a business model for creating value through collaboration between various participants on a network; it very closely resembles a "marketplace."

Sangeet Paul Choudary, in his book *Platform Revolution,*[16] describes the shift in corporate business models from what he calls "pipes" (linear business models) to "platforms" (networked business models). Before the digital revolution, firms created goods and services, which they pushed and sold to customers. The flow was linear, like oil pipes connecting production upstream to consumption downstream. Unlike pipes, business platforms disrupt the clear-cut demarcation between producer and consumer by enabling users to create as well as consume value. Platforms need a different infrastructure than pipes. For example, they need specialized hardware and software that enables and facilitates interactions between the firm, business partners, consumers, and other stakeholders, or what is collectively referred to as the "platform ecosystem." Because platforms resemble marketplaces, where demand and supply must balance, they need to attract sellers and buyers, or providers and consumers, in sufficient numbers in order for transactions to occur; which is often a challenge akin to a catch-22 situation.

Take, for example, a company like Uber. For consumers to be motivated to download the app and order a ride, Uber must ensure that there are enough drivers on its platform. But for drivers to be interested in joining, they need to know that there are enough consumers looking a ride! For this reason, creating a "digital marketplace" (which is another way of describing a business platform) requires considerable initial investment: suppliers have to be incentivized to join even while consumers are still absent, until a critical mass of supply attracts demand and market dynamics kick in.

What appeals to investors who pour millions of dollars into creating a critical mass of supply in a platform is that, once the liquidity of the platform reaches that phase transition point of market dynamics, one gets the so-called network effects. The platform suddenly becomes very attractive to many more suppliers and consumers, which leads to the number and the value of marketplace transactions scaling exponentially. Marketplaces have existed for millennia, of course, but what is different nowadays is that *companies are becoming like marketplaces*, a transformation facilitated by software, powerful computer servers on the cloud, computer networks, and data. And this is where computer platforms and business platforms come together and often merge. In doing so, they shape the nature, playbook, and dynamics of the digital economy with significant repercussions for innovation and democracy.

During the first era of the Internet, computer platforms were "open" and controlled by the Internet community. Developers were free to build whatever applications they wanted and have them run on those open platforms. It was a world of direct democracy not dissimilar to ancient Athens. But by the mid-2000s the Internet had evolved into "web 2.0," which included social media features and advanced search engines. Web 2.0 was touted as extending the democratic franchise of the Internet beyond the software developing cognoscenti to include everyone with access to a computer and the ability to post a photo on their social media feed. But instead of more democracy a new oligarchy of five big players emerged: Facebook, Amazon, Microsoft, Google, and Apple, collectively known as "FAMGA." The business models of those digital behemoths fuse the ideas of computer and business platforms into an unbeatable combination that is generating enormous investor value.

For instance, Google has developed the operating system Android (a computer platform), which enables a three-way marketplace (a business platform) connecting software developers (who write apps), hardware developers (the companies that manufacture smartphones and wearables), and users. Indeed, the explosion of smartphone use has been one of the main drivers for the digital transformation of companies across every industry, since it has radically changed consumer behavior and allowed for highly personalized consumer experiences. This explosion, however, has had several side effects. The first is that the "open" and decentralized computer platforms of the first era of the Internet have been replaced by closed, centralized services offered by big tech companies. The centralization of computer platforms means that software developers and entrepreneurs must abide by rules set by the big tech's strategic agendas—rules that can change at any time, unpredictably, serving the interests not of the "many" but of the "few," which usually means the big tech's shareholders. In other words, the competitive field on web 2.0 is no longer level but profoundly skewed toward the owners of incumbent, closed computer platforms.

PLATFORM CAPITALISM

The implications of the so-called platform capitalism for working people are considerable. "Digital transformation" pushes traditional companies to radically change their linear, pipelike business models into platforms and imitate FAMGA success. These models do not just reinvent how companies engage with their customers; they also change how work is done. The platform model demands a marketplace for resourcing and paying for work, linking demand with supply in a dynamic and cost-efficient way. As businesses undergo digital transformation, they discover that they need far fewer full-time employees, as they can now engage the necessary skills from a plurality of other sources, such as talent platforms, consultants, contractors, part-time workers, and sometimes volunteers.[17] Jobs are deconstructed into tasks, and tasks are either automated or distributed to various suppliers of work. This work deconstruction and task redistribution is orchestrated using specialized software that matches work with skills[18] using AI algorithms.

As digital technologies dissolve the traditional organization of the firm and deconstruct jobs into tasks executed by a mix of humans and algorithms, anxiety is growing about the "gig economy"—where work is temporary, skills based, and on demand. The numbers of independent workers are exploding across the developed world. A study by the Freelancers Union found that 53 million Americans, or about 34% of the total workforce, are independent workers, a number expected to rise to 50% by 2020.[19] In Europe there is a similar picture emerging, with about 40% of EU citizens working in the "irregular" labor market, according to the European Commission.[20] A McKinsey study[21] looked in depth at the numbers, as well as at how independent workers feel about work, and confirmed the tension between the positive aspect of going independent, which is flexibility, and the negative, which is income insecurity. Whether you are a driver for Uber, or a financial analyst contracted via Upwork, or a software developer microtasking in some company's development framework, your livelihood depends on the cyclical fluctuations of demand for your skills, your rating, and your ability to market yourself effectively against the competition. Moreover, residual profits from your participation never reach you but go straight to the owners of the platform, who are not currently obliged to provide participants with the protections associated with employment.[22] Add the threat of automation from AI, and the result is a survival-of-the-fittest scenario coupled with a race to the bottom for rewards and benefits.

As many companies around the world begin to transform their organizational models to become more agile, the "platforming" of full-time employees is becoming a top management agenda item. Take Haier, the Chinese manufacturing giant, which is transforming into a platform for entrepreneurship by encouraging its employees to become self-governing entrepreneurs. Zhang Ruimin, the CEO of Haier, sees this transformation as the only path that can secure a future for his company.[23] He is removing traditional leadership and replacing it with worker autonomy. Such bold moves may benefit risk-takers but are a threat to workers who value security and stability more. In many cases the replacement of human hierarchies with software systems that orchestrate work clashes with human nature, as in the case of Zappos, the digital shoe and clothing shop currently owned by Amazon. The company employs around 1,500 people and has a billion dollar annual turnover. It has implemented a self-organization methodology invented by a software engineer[24] known as "holacracy." Instead of pyramidal hierarchies, holacracies are organized around "circles"; each circle can encompass a traditional function (such as marketing) as well as other "subcircles" that focus on specific projects or tasks. No one prevents workers from freely moving across subcircles in order to achieve their goals, because there are no managers to stand in their way. Instead, software enables collaboration and the performance of individuals and teams, while "tactical meetings" allow for employees to provide feedback about how things are working in a tightly circumscribed format. Yet despite the flexibility and efficiency holacracy promises, it's been criticized for not taking into account the emotional needs of workers, and reducing humans into "programs" that must run on the operating system of digital capitalism. Similarly, many Uber drivers often report[25] feeling less like humans and more like robots, manipulated by the algorithm in the app, telling them exactly what to do. When jobs are deconstructed into tasks, and those tasks are then automated using AI systems, there is a risk that the human workers of today will be completely transplanted by intelligent machines.

DIGITIZATION, DIGITALIZATION, AND THE AUTOMATION OF WORK

The displacement effect of technological disruption is always easier to predict than the compensation effect. That is because most new technologies are invented in order to generate cost efficiencies in the present, which is generally quite well understood. On the other hand, predicting how these technologies will generate a need for "new work" is a lot harder, as there are too many unknowns to calculate. This explains why most of the economic analysis on the forthcoming Fourth Industrial Revolution has concentrated on jobs lost rather than jobs gained. Easier does not, however, imply "easy": predicting which jobs will be automated—or "displaced"—by intelligent machines, and how quickly this will happen, remains a difficult task wrought with many assumptions and oversimplifications and dependent on the methodological approach economists adopt in making their predictions. Many economist think that the displacement impact of automation will range from considerable to devastating, as, for example, in the famous 2013 paper by Oxford economists Frey and Osborne[26] that predicted a loss of 47% of jobs in the United States by 2023. Using a related methodology, McKinsey put the same number at 45%, while the World Bank has estimated that 57% of jobs in the OECD member states will be automated over the next two decades.[27]

Work automation not only reduces the number of jobs available but also negatively impacts wages. Research by economists Daron Acemoglu and Pascual Restrepo looked into US labor markets and estimated that one more robot per thousand workers reduces the employment-to-population ratio by up to 0.34% and wages by up to 0.5%.[28] This seems peculiar: technological innovation has historically boosted human productivity, and one would expect that robots and AI would do the same now and that wages will rise. Indeed, a report by the consultancy firm Bain[29] estimated that productivity would rise by 30% across all industries because of automation. And yet, if one combines these two findings, it looks as if the providers of work will not be sharing in the new bounty of automation. Wages will be depressed *despite an increase in productivity*. This result is even more puzzling if one factors in demographics: the number of workers will decrease in the next decades both in the West and in China, Korea, and Japan. And yet, the decrease in the supply of labor does not seem to lift wages either.

To understand this worrying trend, it is important to distinguish between *digitization* and *digitalization* and then proceed with analyzing how digitalization shifts the balance of contribution in economic value creation from human workers to software systems, that is, from labor to capital. *Digitization* is the process of rendering a physical object in a digital form as zeroes and ones. Scanning a paper document and recording a voice and storing the recording as a digital file are examples of digitization. *Digitalization* replaces methods of sharing information in a process with computer instruction code. For example, a process whereby a clerk would review a number of paper documents in multiple binders and then walk to the office of her supervisor, three floors up, to wait outside his door and ask for approval, could be digitalized using computer programs. One often needs the digitization of physical objects to happen before proceeding with the digitalization of a process. In the example of the clerical worker, digitizing paper documents and binders into digital files and folders provides the opportunity to develop computer code that digitalizes the process of review and approval using telecommunications instead of physically climbing stairs and waiting for doors to open. Digitalization "automates" tasks in a process, and often the whole process, reducing cost, increasing efficiency, and boosting productivity. Process tasks need not be trivial. For instance the digitization of radiology scans allows for a computer, instead of a medical expert, to perform diagnosis—that is, it allows for the digitalization of the diagnostic process. Digitalization's other consequence is that it requires humans to acquire new skills in order to compete in the new, "automated" world. In the example of the clerk, the ability to access a computer system and manipulate digital files, use email, and so on are the new skills she needs to retain her job. The fact that the process no longer needs the clerk's physical presence in the workplace is another profound consequence of process digitalization. Workers and workplaces can be disentangled. The clerk can work remotely and perform the new process. She can be anywhere in the world.

Given the interplay of digitization and digitalization, robots do not replace humans directly; it is the process of digitalization that does so in very unpredictable ways. The roboticist Rodney Brooks gives the example of the human toll collector.[30] Developing a robot that would do exactly the same job is hard and inefficient. The dexterity required to reach out and meet the outstretched arm of a driver to collect coins and notes in a windy environment is hugely challenging for a robot with present-day technology. Nevertheless, *the human toll worker can be replaced by digitalizing the toll-paying process*. For this to happen a number of innovations have to occur. The car must acquire transponder capabilities and transmit digital information regarding ownership; credit cards have to be digitized so they can be charged without the need of physical contact; wages need to be credited digitally into a bank account connected to a credit card, and so forth. Digitalization is therefore an emergent phenomenon that is virtually impossible to predict, because it is the unpredictable *combination of technological innovations*.

WHEN HUMANS ARE NO LONGER NEEDED

Automation technologies are advancing apace. There can be no doubt that they will impact the current economic model of growth in a profound and unprecedented way. It is not necessary for AI to reach, or overtake, "human-level" intelligence, or for the robots to become as "dexterous" as human beings, for human work to be replaced almost entirely by machines. Digitalization does not replace human workers by simulating how humans perform work tasks; it does so by reconfiguring how work is done in a process. This reconfiguration can be significant and replace many tasks that we currently regard as uniquely human.[31] Take, for example, how machine-learning systems diagnose cancer by looking into medical images. They do not do so by emulating human experts. Instead, they do so by scanning thousands of medical images and building their own internal ways of drawing inferences from data correlations. They do not need to understand what they do or why they do it. Their "intelligence" is different from ours—it has no semblance of consciousness or self-awareness—but none of that is important. What is important is that intelligent machines can do diagnosis better than human experts, do it faster, and do it at a scale. Also important to note is that a machine is a capital asset. In hospital systems we have capital assets, such as buildings and equipment, and human resources, such as doctors, nurses, orderlies, janitors, and so on. Given what cognitive technologies are capable of, it is not too hard to imagine a hospital system that is highly digitalized and where markedly fewer human resources are needed. In that future hospital the ratio between capital assets and labor will have shifted significantly in favor of the former. In fact, the future hospital may be completely made of software.

The compensation effect has always created demand for new work during past industrial revolutions. But given the exponential rate by which automation technologies improve themselves, we may be reaching a tipping point in history where most of the new work created by AI will be done by AI! Take, for example, Uber drivers: a software-based platform in combination with satnav systems and the digitization of payments has delivered a high degree of digitalization of the process of finding a taxi when you need it. As a result, many microtasks of the process have been automated: as a passenger I do not need to step out onto the street and wait in hope for a passing cab, a taxi driver does not need to know all the streets and alleys of a city to get me to my destination, I don't have to dig into my pockets for bills and coins to pay the driver, and so forth. This "first wave" of digitalization due to AI has created some new jobs for humans, albeit low-paying ones. Nowadays, almost anyone can be an Uber or a Lyft driver, and many have done exactly that, either as a means of supplementing their income or as their main job. Who doubts, however, that the next wave of the digitalization of the taxi process will be the total elimination of human drivers and their replacement by powerful AIs?

In fact, few people do. According to research by Pew,[32] "most [citizens] believe that increasing automation will have negative consequences for jobs" and "relatively few predict new, better-paying jobs will be created by technological advances." Citizens are clearly aware of the danger that AI-powered automation will destroy their livelihoods, and the data back up their perception. Research on US Department of Labor data by Axios revealed that three-quarters of US jobs created since the 2008–2009 financial crisis pay less than a middle-class income.[33] As the US economy is on a record-breaking streak of adding new jobs into the labor market, professions that were once the backbone of the middle class have been vanishing. This "hollowing out" phenomenon is reflected in the current composition of the workforce, which is concentrated mostly on high-skill, high-wage and low-skill, low-wage jobs. Middle-skill, middle-wage jobs are vanishing. Manufacturing—because of the degree of automation already in place—is a good proxy indicator for the future of work, as technological disruption is already impacting more middle-skill, middle-wage jobs across all industries. According to data by the Federal Reserve, 25% of jobs have disappeared from manufacturing in the last two decades.[34] Given the exponential rate of technological change, we should expect a much higher percentage of hollowing out of the labor market in a much shorter time frame.

The Fourth Industrial Revolution could lead to the massive digitalization of business processes across every industry. If that happens, human work—as we know it—could become unrecognized. Full-time jobs would become a rarity, and most of us would be working part-time for a wide variety of clients. Meanwhile, society would be reaping handsome dividends from this new transformation of the economy. Machines will be producing goods at a very low cost in fully automated factories—or indeed much closer, in our homes, using 3D printing. Digital assistants powered by AI will be our doctors, financial advisors, and personal agents looking after every aspect of our lives. Innovation will be accelerated across every industry. We can imagine AI systems used by scientists to synthesize new materials that mimic plants and make superefficient solar panels. Solving our planet's energy and environmental problems at a stroke, we can proceed by using AI to optimize food production and every other process too, so that human productivity rises exponentially. In such a scenario we could, theoretically, fulfill most of our material needs without the need to work and, in effect, return to an era of economic abundance. But there is a caveat.

Unlike our prehistoric ancestors, we now live in highly complex societies with sophisticated power structures and institutions, where wealth redistribution is largely controlled by governments and whoever has the most influence on their decisions. Therefore, a key political question about the Fourth Industrial Revolution is *how governments will enact policies that encourage the widest possible sharing of the economic bounty that intelligent machines will deliver to the economy*. The risk of not sharing the spoils of the AI economy in a democratic way is rather obvious: ordinary citizens whose only way to acquire income and wealth is through work would sink into a spiraling abyss of abject poverty. Such an outcome would most certainly destroy the last vestiges of trust in liberal democracy. Democratic governments would therefore need to step in and manage the impact of automation on work. They would also need to deal with many other risks and challenges that AI brings, including, the ethics of AI algorithms and the use of personal data in business and government. Governments would need to adopt AI systems in their processes, in order to improve their own performance and deliver better services to citizens. But how far can we—or should we—go with the automation of government? And what should the role of government be in an automated future? Given the potential of AI to automate many cognitive processes at very low cost, as well as its superhuman power for prediction and strategic planning once it processes vast amounts of data, can we imagine a fully automated government, one that is run by intelligent machines rather than human civil servants and politicians? Does it make sense to relinquish control of our economies to intelligent machines with full autonomy? The next chapter will examine in more detail such questions, as well as the different approaches between liberal democracies and authoritarian regimes in shaping the future of AI.

4

MACHINES AT THE HELM

For most citizens in liberal democracies, life is dependent on the provision of good-quality public services in education, health care, retirement, policing, taxation, and the courts—to name but a few. These valuable public services are becoming extremely costly over time, contributing to the swelling of government deficits and the ballooning of national debts. Bad demographics make this trend worse, as many liberal democracies are trapped in a vicious circle of ever-diminishing tax receipts, due to the scarcity of young, taxable workers, and an aging population demanding more public services. This situation is clearly unsustainable. Artificial intelligence is a technology that could help alleviate some of these imbalances—for example, by optimizing the distribution of scarce public resources, improving decision-making based on intelligent predictions, and preventing tax fraud. Shouldn't we therefore welcome the computer automation of government and public services using AI?

The relationship between computers and governments goes back a long way. In *The Government Machine*,[1] historian Jon Agar argues that the ideological roots of computers are found in public administration. According to Agar, the mechanization of government started in the late eighteenth century when the public administration of the United Kingdom, trusted with running a global empire, invested in efficient operations, as well as in the collection and processing of information from across the world. In a liberal system of government, where civil servants run everyday government affairs, the computer is a reflection of a technocratic vision for efficiency and process management. As Agar says, "The general purpose computer is the apotheosis of the civil service."[2] Until recently, computer systems were acting as the "peripheral nervous system" of government organizations: automating some processes, collecting data, and using simple interfaces to serve citizens over the web. Extending the brain metaphor, the "central nervous system" of government has remained the cabal of top-level human administrators; the departmental directors, directors general, and ministers. For it is at that level where the ultimate responsibility for decision-making and action still resides. Nevertheless, complex government decisions require not just data and information fed upward by administration processes but expert advice too. Indeed, the rise of expert advisors in public administration—such as statisticians, health scientists, economists, and so on—is absolutely necessary for "evidence-based" politics, or what the German philosopher Jürgen Habermas calls the "scientization of politics."

Enter AI, which can potentially automate the central nervous system of government, as well, and deliver an efficient, mechanized "organizational brain" that can make complex decisions autonomously by accessing vast amounts of diverse knowledge and data. By replacing human decision makers in public administration with informational processes controlled by AI algorithms, one could get the perfect government: impartial, efficient, and effective. But what would be the consequences of transforming the metaphorical "government machine" into a reality?

AUTOMATING THE GOVERNMENT

Claiming that the combination of AI and big data is a better way to run government decision-making and services may be both sensible *and* undesirable at the same time. By extracting human empathy from government, and transforming human subjectivity into mechanical objectivity, we may end up with something worse than we bargained for. Virginia Eubanks, in her book *Automating Inequality*,[3] presents a number of case studies in the use of automation and algorithms by public service agencies in the United States. In Indiana, an automated benefits system categorized its own errors as "failure to cooperate," and, as a result, wrongful denials of food stamps soared from 1.5% to 12.2%.[4] More worryingly, Eubanks notes, in the name of efficiency and fraud reduction automation replaced human caseworkers who would exercise compassion and common sense in helping vulnerable people. Many other examples point to the unintended consequences of replacing humans with algorithms. COMPAS is an AI algorithm widely used in the United States to guide sentencing by predicting the likelihood of a criminal reoffending. In May 2016 the US organization ProPublica reported that the system predicts that black defendants pose a higher risk of recidivism than they actually do.[5] Another similar case is the algorithm PredPol, used by police in several US states to predict where crimes will take place. In 2016 the Human Rights Data Analysis Group found that the system led the police to unfairly target certain neighborhoods with a higher proportion of people from racial minorities.[6]

The problem with automation is linked to the data-driven nature of machine learning. Algorithms "learn" by self-adjusting internal probabilistic weights through thousands of successive iterations with data sets. But data may have inherent biases. Safiya Umoja Noble, in her book *Algorithms of Oppression*,[7] recounts her experience when, looking for inspiration in how to entertain her preteen stepdaughter, she searched for "black girls" in Google. To her horror, instead of eliciting information that might be of interest to this demographic, the engine produced results awash with pornography. Algorithms that feed on user citations tend to pose a major threat to the human rights of marginalized groups. Black teenage boys, Noble notes, are to be found next to criminal background check products. Gender equality is also affected. The word "professor" returns almost exclusively white males, as does the word "CEO." Because of that, Google's online advertising, which feeds from search results, shows high-income jobs more often to men rather than to women, thus perpetuating the gender imbalance. Bias in data reinforces societal inequalities and prejudices. Moreover, a weakness of deep neural networks is that they cannot explain the reason for their outputs. The combination of data bias and algorithmic inexplicability can be highly problematic when AI systems have an impact on citizens' lives. From a classic liberal perspective, it is politically intolerable as it alienates citizens from the state and transforms the latter into an authoritarian and oppressive machine.

AI algorithms can also shape public opinion, impact electoral results, and be a direct threat to the liberal system of government. Algorithms used in social media platforms optimize the content citizens receive by "profiling" them. By doing so, they reinforce our biases while excluding us from debates, information, and dialogues that may lie outside our narrow interests or be at odds with our political ideology. With two-thirds of American adults getting their news from social media,[8] this "algorithmic segregation" played a crucial role during the 2016 US election. As social media platforms gain enormous influence in shaping public opinion, all kinds of possibilities are open for nefarious interference in national politics by powerful interest groups or hostile countries. Cambridge Analytica, a company based in the United Kingdom, used data from Facebook and its own AI algorithms to personalize messages and "microtarget" voters in the US election, to great effect.[9] The fact that they were exposed by the media and driven to bankruptcy does not minimize the fact that what they did is what digital advertising agencies do every day. We live in a world where AI algorithms modulate content around our personal views and prejudices, constantly reinforcing them and rarely challenging them. That's how advertising works, and advertising happens to be at the heart of business models in content-based platforms such as Google, Facebook, and Twitter. The undesirable by-effect of AI-powered personalization in politics is polarization, wherein everyone becomes convinced that they are right while all others of a different opinion are wrong. This leads to a breakdown of consensus within our societies and is contributing to the rise of illiberal populism and the establishment of Internet echo chambers instead of a democratic agora.

Nevertheless, given the huge benefits of AI, we must find a way to embed this revolutionary technology in our system of government without jeopardizing liberal values. How we achieve this is fundamentally a question of ethics, and many initiatives across the Western world are grappling with issues such as algorithmic explainability, data bias, reinforced social exclusion, and the limits of systems autonomy. The debate on AI ethics is discovering how complex it is to define universal standards for AI. The professional association of electrical and electronics engineers—the IEEE—launched a global initiative on Ethics of Autonomous and Intelligent Systems in order to develop consensus "on standards and solutions, certifications and codes of conduct . . . for ethical implementation of intelligent technologies."[10] Their approach decided to examine ethics not just from a "Western," Judeo-Christian perspective but also from other cultural perspectives, such as African and Chinese. IEEE's work is arguably the most comprehensive compendium to date on how to mitigate the societal risks of intelligent systems. But, as the authors aimed for a standard for "common good," they discovered that the idea was inconsistent with a pluralistic society.[11] Efforts to force a specific notion of common good would inevitably violate the freedom of those who do not share this goal and inevitably lead to paternalism, tyranny, and oppression. For example, the individualism of European and American societies often clashes with the communitarian values of African societies. The authors of the IEEE report, in agreeing that human values differ across cultures, highlight one of the most important aspects in the geopolitics of AI. Given differences in human cultures, we may indeed end up with different AIs in the future—for instance, African AIs that prime collaboration versus European and American AIs that prime competition.

Meanwhile, and as debates on AI ethics continue, regulation is already being put in place on both sides of the Atlantic in order to limit the discriminatory and biased nature of machine learning, as well as the tremendous influence that AI systems are already exerting on democratic processes. The European Union's General Data Protection Regulation (GDPR), launched in 2018, requires that AI systems be capable of explaining their logic and demands transparency in how personal data are manipulated. In a similar vein the State of California passed, on June 28, 2018, the California Consumer Privacy Act, which protects the privacy rights of consumers within the state. At a federal level, in 2019 Senators Cory Booker (D-NJ) and Ron Wyden (D-OR), with a House equivalent sponsored by Rep. Yvette Clarke (D-NY), proposed the Algorithmic Accountability Act.[12] The act asks the Federal Trade Commission to create rules for evaluating "highly sensitive" automated systems. Companies would have to assess whether the algorithms powering these tools are biased or discriminatory, as well as whether they pose a privacy or security risk to consumers.[13] And while we grapple with the many challenges that AI poses for our liberal institutions, democratic system of government, and values, a different story is developing in communist China.

AI AND COMMUNISM

In the late twentieth century the command economies of communist countries in Eastern Europe and the Soviet Union imploded because they were unable to compete with free and open markets. That happened because the latter were able to allocate capital more efficiently, by quickly discovering prices on the basis of supply and demand, and thus produce better products and services. In the planned economies of communist countries it was not possible to use the markets to discover prices. Instead, central planners were burdened with solving the so-called socialist calculation problem: data about production capacity and consumer demand had to be collected using surveys, and an enormous number of mostly manual calculations had to take place in order to discover prices that reflected the true value of economic exchanges. Those prices were then used to set production targets. The free market economist Ludwig von Mises thought that the socialist calculation problem was a fool's errand.[14] He argued that the complexity of an economy is so great that only a free market mechanism can discover true prices. Without true prices any economy is doomed to failure, as it would waste valuable resources in producing too many useless goods. For von Mises free markets are more efficient than central planning because they are natural calculators of prices.

Nevertheless, the dream of a centrally planned economy never died.[15] The Polish economist and father of market socialism, Oskar Lange (1904–1964), claimed that solving the socialist calculation problem was theoretically feasible. He proposed that an economy could be described as a series of simultaneous equations. Leonid Kantorovich (1912–1986), a Soviet mathematician who invented linear programming and won the Nobel Prize in economics in 1975, tried to solve Lange's equations. He spent six years gathering data and running calculations to optimize Soviet steel production. When Kantorovich was trying to solve Lange's equations, the Soviet Union produced around 12 million types of products. Kantorovich failed not because he did not get good results but because his results would always come too late to be useful. Arguably, his failure was due to the slowness of his calculations, restricted by the computational power that was available to him at the time.

The 1970s saw one more effort by a socialist government to solve the socialist calculation problem, in Chile. As soon as Salvador Allende became the first Marxist president of Chile in 1970, he embarked on a massive program of nationalization and collectivization. His vision was to transform his country into a socialist utopia and succeed where the Soviet Union and Eastern Europe were failing by creating an egalitarian society where the economy was managed scientifically from the center for the benefit of every citizen. To his aid came the British cybernetician Stafford Beer, a pioneer in organizational cybernetics. Beer designed and helped build Cybersyn, a cybernetic system that would be the "nervous system" of the new Chilean socialist state.[16] The system ran on a mainframe computer and collected data from factories via 500 telex machines. The data were then fed into a simulator that ran various mathematical models in order to predict possible outcomes of various decisions and policies. A central control room was built in Santiago, the Chilean capital, where the managers of the socialist Chilean economy could meet and control production levels across the country. This was supposed to be the future, the end of history, and the best way that the world ought to be run. Except Cybersyn was never put to the test. Following a military coup in 1973 the Allende government was ousted, and Chile became a military dictatorship under General Pinochet. Cybersyn was mothballed, and Milton Friedman, the University of Chicago guru of free market economics, arrived in Chile to nudge the country in a completely different economic direction.

Fast forward to today when machine-learning algorithms are capable of discovering correlations in massive, unstructured, and disparate data sets and of using them to make predictions and classifications that far surpass the capability of any human. Not even in their wildest dreams could Lange, Kantorovich, Beer, and Allende have imagined the current state of computer technology. Could AI, big data, and supercomputers provide the technological means to solve the socialist calculation problem? The Chinese communists definitely think so.

In May 2018, Professor Feng Xiang, one of China's most prominent legal scholars, published an op-ed in the *Washington Post* entitled "AI Will Spell the End of Capitalism."[17] He argued that in China's socialist market economy AI could rationally allocate resources through big data analysis and robust feedback loops—practically echoing the cybernetic model of a socialist economy that was imagined in Cybersyn. Given that work automation will cause mass unemployment and demand for universal welfare, Xiang suggests that AI and big data should be nationalized. Private companies would coexist in this cybernetic communist scenario but be closely monitored by the state and under social control. Instead of corporate bosses serving the needs of shareholders, as in capitalism, Chinese business leaders would serve the needs of the worker-citizens. Xiang's op-ed was a window into how the political elite of China thinks of the future of the Chinese economy and political system in the age of intelligent machines. After all, the ultimate goal of communism is the elimination of wage labor. Automation of work should be celebrated and embraced. Kantorovich and Lange are being revisited in earnest, as evidenced in the work of Chinese economists Binbin Wang and Xiaoyan Li who demonstrated how a combination of machine learning and low-cost sensors could optimize production in real time, as well as personalize it to the needs of citizens.[18]

However, for this proposition to become reality, the Chinese government must have unimpeded access to citizen data. The AI solution to the socialist calculation problem requires citizen surveillance. China's social credit system should be seen as part of this grand vision for the future of communism, and not simply as a way for the Chinese government to spy on citizen's lives, although the latter is clearly one of its goals. For instance, the social credit system is mostly developed in the Xinjiang Province, where it is used to monitor and control the Uighur population; many citizens deemed unsafe are shut out of everyday life or sent to reeducation centers in the province.[19] To the rest of China, the social credit system is making citizen surveillance into an—almost amusing—game whereby citizens are incentivized to behave in specific ways and, if successful, get rewards. The reward scheme is based on a point system with every citizen starting off with 100 points. Citizens can earn bonus points up to the value of 200 by carrying out "good deeds," such as doing charity work, recycling rubbish, or donating blood. Having a high social credit score opens up doors for a better job, access to health and education, travel and leisure, and even getting matched to the right partner. But citizens can also loose points by acts such as not showing up to a restaurant without having cancelled the reservation, cheating in online games, leaving false product reviews, and jaywalking. Not having enough social credit points carries costs, such as citizens getting banned from taking flights and boarding trains. In early 2017, the country's Supreme People's Court announced during a press conference that 6.15 million Chinese citizens had been banned from taking flights for social misdeeds. For Chinese communists, inspired by Confucian ideals wherein the whole is prioritized over the individual, such top-down interventions are both ethical and legitimate. Social trust must be preserved at any cost, even if a great number of citizens get penalized forever. As President Xi Jinping has stated, rather bluntly, "once untrustworthy, always restricted."[20]

The social credit system is also considered by the Chinese government as an alternative to the democratic process of elections. By scraping citizen data from social media feeds and other digital channels and analyzing them with AI algorithms, the communist rulers of China can "listen" to their people and understand what they want, think, and feel. This idea of replacing democracy with computer analysis echoes a science fiction short story written in 1955 by Isaac Asimov.[21] In that story a single citizen, selected to represent an entire population, responded to questions generated by a computer named Multivac. The machine took this data and calculated the results of an election so that it never needed to happen. Asimov's story was set in Bloomington, Indiana, but an approximation of Multivac is being built today in China.[22]

SHARING POWER WITH AIS

The contrast between communist China and liberal democracies in the debate around algorithmic accountability and data privacy could not be starker. What our liberal values find shocking and abhorrent—such as government systems dictating outcomes for citizens without the power to appeal—the Chinese implement with enthusiastic zeal. Nevertheless, both liberal democracies and Chinese communists aspire to use AI and data in order to improve human lives, protect Earth, and give more opportunities to future generations. Therefore, we have a common interest in fully understanding and wisely deciding the degree and level of AI automation that we should allow in our societies. So let us imagine an extreme future scenario in which intelligent machines have replaced humans at the helm of government, automating most decision-making processes, crunching massive data in fractions of the time that it would take humans to do, solving the socialist calculation problem, predicting crime offenses and loan defaults, and controlling every aspect of citizen behavior in general.

Let us also imagine—this time in a liberal democracy context—that we have solved the problem of data bias and algorithmic oppression, and that our algorithms are now ethical, they can explain their reasoning, and there is an appeals process in place to protect citizens from algorithmic bias. The running of government is now largely outsourced to those ethical intelligent systems—our new agents—that "know" everything better than us because they can learn everything faster and can be everywhere, anytime, even while we sleep. The machines are making the big and small decisions of running a country; they distribute resources according to one's needs, enforce the laws, and defend the realm. Perhaps there is some kind of democratic oversight of this algorithmically based government; let's say there is an elected parliament that regularly checks the performance of the algorithms and has the power to order their removal or improvement. Meanwhile, we citizens can enjoy our lives *all watched over by machines of loving grace*,[23] without worrying about corrupt politicians, special interest groups, corporations, bankers, or boom-and-bust cycles. The machines are objective, have no emotions, have no vested interests, and will make sure none of the past shortcomings of politics will ever harm us. The machines will protect and defend liberal values and ensure the protection of citizen and human rights. Their interests will align perfectly with the interests of the many, and the agent-principal problem of representational governance will be all but solved. From a liberal perspective, such a fully automated future may seem ideal. From an economic perspective the AIs would deliver maximum efficiency. It all sounds great, but would such a future be desirable?

The key to exploring the desirability of a fully automated government lies in the concept of "autonomy," that is, in how the responsibility for an action, or a decision, is shared between a human operator and a machine. The degree of autonomy also implies the degree of influence that humans have over the application and further development and evolution of AI systems. In the futuristic scenario under discussion, the machines are fully autonomous and there is no need for a human to be involved. The autonomy level just before that is when a machine is largely autonomous but a human must intervene in a preselected set of decisions that usually have to do with extreme events and circumstances; that is, the human is a "supervisor" of last resort. Let's examine and compare those two extreme scenarios of autonomy, not in politics, but in real cases wherein the outcome was either life or death.

To optimize aerodynamics and fuel efficiency, Boeing designed its new aircraft 737 Max 8 so that it incorporated a fully autonomous system that took independent action to correct pitch and prevent stalls when the plane climbed too steeply. Unfortunately, the system—called the Maneuvering Characteristics Augmentation System (MCAS)—was shown to have caused the crashing of two flights in 2019, Ethiopian Airlines Flight 302 in Ethiopia and Lion Air Flight 610 in Indonesia, with a combined loss of 346 passengers and crew.[24] In both cases the pilots failed to counteract actions by the autonomous system. They fought the automated system trying to pull the nose back up but did not succeed. By making the system fully autonomous Boeing had virtually cut the pilots out of the aviation process. The full automation of flight has been an aspiration for many in the industry. There used to be a saying that in the future airplanes would need only one pilot and a dog in order to fly. The pilot would feed the dog and the dog would make sure that the pilot did not fly the plane. MCAS was the dog.

The second case, where the intelligent system was not fully autonomous but allowed for human supervision, had a happier ending. On September 26, 1983, just a few minutes after midnight, Lt. Col. Stanislav Petrov was sitting in the commander's chair inside a secret nuclear missile launch bunker outside Moscow when alarms went off: satellite data indicated that the United States had launched nuclear missiles. Computers were certain that the Soviet Union was under attack. They suggested an immediate retaliation in kind. It was a testament to common sense and moral integrity that Lt. Col. Petrov decided to act against the advice of computers that night. He did not push the launch button. Had he done otherwise, we would probably not be alive today. In fact, if the Soviet Union had developed fully autonomous AI systems for its nuclear arsenal that did not need the supervision of officers like Petrov, mutual nuclear annihilation would have probably happened already. The Rand Corporation examined in detail such cases in its Security 2040 project[25] and found that fully autonomous AI would significantly increase the risk of tensions escalating into a full-blown nuclear war.[26]

Nevertheless, we must not regard the Lt. Col. Petrov's case as conclusive evidence that humans as supervisors of autonomous systems are the better approach to automation. In fact, having a human in the loop can often be catastrophic. The fatalities recorded so far with driverless cars have been mostly due to the not yet fully autonomous car handing control over to the human driver with little time for the human to react. It seems that we humans are not good at taking considered action by overriding complex autonomous systems in time-critical situations. When working with such systems, we get "automation complacency" and trust the machines too much, until it is too late. It is notable that insurers considering risk in driverless cars are regarding this uncertain interface between human supervision and system autonomy as highly risky.

Faced with the dilemma of full versus partial automation, we seem to be caught between Scylla and Charybdis. Both options could lead to catastrophic outcomes. So what should we do to capture the opportunity of AI while mitigating the risk? Perhaps, and in order to extricate ourselves from the conundrum, we need to revisit what AI means. Our current idea of AI is one where humans are viewed in juxtaposition to algorithms. Artificial intelligence systems are like aliens invading our world, and we are now forced to somehow find a way to accommodate them and hopefully share power and responsibility with them. So, maybe, we got this whole thing wrong. Maybe we need to rethink AI, not as something that we must try to adopt by reducing the risk of bad outcomes through regulation and safety checks but as something that has to be embedded in human systems in a symbiotic way, and only when it maximizes our personal and collective goals. Such an approach would require a more dynamic relationship between AI and humans and more decentralized, bottom-up, and democratic processes that empower citizens to influence the further development and evolution of AI algorithms. In a later chapter I will propose and discuss a theoretical framework that can help us rethink AI in those democratic and human-centric terms. But before delving into the details of this framework, let us take a high-altitude view and examine how AI and its accompanying dilemmas are reshaping the world order, as well as what options are available to liberal democracies in order to absorb the impact of automation while remaining true to liberal values.

5

A NEW WORLD ORDER

Since the end of the Cold War, most political theorists have regarded liberal democracy as the only path to economic success. Empirical evidence from the collapse of the Soviet Union and the Eastern European economies has supported a profound dichotomy between countries that repressed their people and became poor and those that liberated them to unleash the forces of creativity, innovation, and economic growth. Artificial intelligence is now disrupting the economic superiority of liberalism over authoritarianism. It is now possible for authoritarian regimes to use AI in order to restrict individual liberty while allowing for economic development. As we saw in the previous chapter, China is using AI to do exactly that. By deploying a combination of citizen surveillance using the social credit system and selective censorship via the so-called Great Firewall, China is constructing a digital authoritarian state. The social credit system centralizes citizen data for predictive analysis. The government-controlled firewall blocks citizen discussions as well as content that the Communist Party decides to be potentially detrimental to its absolute authority, including the sites of many Western nongovernmental organizations (NGOs) such as Amnesty International, as well as YouTube, Twitter, Google, BBC, the *New York Times*, and the *Wall Street Journal*.[1] At the same time, China is allowing information for productive activities to flow freely so that it may boost its economic development.

Because of China's economic success, digital authoritarianism by an all-seeing state is seeking imitators and is spreading around the world. Beijing is exporting its model to Thailand and Vietnam. According to news reports, Chinese companies are providing support for government censors in Sri Lanka and have supplied surveillance or censorship equipment to Ethiopia, Iran, Russia, Zambia, and Zimbabwe.[2] YITU, a Chinese AI firm, has sold wearable cameras with facial-recognition technology to Malaysian law enforcement.[3] The Russian government is establishing digital authoritarianism with great enthusiasm, too, by ensuring that domestic Internet traffic does not get routed via computers abroad.[4] Blocked websites include LinkedIn, the messaging service Telegram, and sites showing how to circumvent Internet censorship.[5] In 2019, Vladimir Putin signed a new law that punishes individuals and media that post online content that shows "disrespect" for the president or the government. Typically for an authoritarian regime, the new law is meant to defend the Russian state from "fake news," as all political criticism of Putin's kleptocratic government is deemed to be lies.[6]

Digital authoritarianism using AI provides dictatorships with new, and extremely powerful, tools for controlling citizen behaviors and thoughts. Unlike previous retroactive censorship methods, AI can use big data to exercise predictive control. This is not different from consumer targeting taking place in the West where an AI system can predict what each of us may want to buy *before we have actually thought of buying it.* The difference is that in the West, companies such as Google, Amazon, and Facebook are only allowed access to specific personal data from some accounts and devices. Regulation in liberal democracies is restricting access to our personal data by private companies in order to protect our privacy and defend us from surveillance. But this is not the case in dictatorships: they can enact laws that centralize and combine citizen data from multiple sources including tax returns, medical records, criminal records, sexual-health clinics, bank statements, genetic screenings, biometrics, geographical location, information from family and friends, consumer habits, and money transactions.[7] Authoritarian regimes can then use this centralized pool of data to train AI systems for the ultimate social control. Knowing that the state can anticipate what you might do, or contemplate, leads to changing not just your behavior but your way of thinking as well. Psychological research has shown that making people adopt certain behaviors can lead to changes in attitude and self-reinforcing habits.[8] Nudging such behaviors using predictive AI algorithms is how China's social credit system is enforcing upon its citizens the most effective disciplinary mechanism ever invented: self-censorship.

A DIGITAL IRON CURTAIN

Artificial intelligence is bringing not just a new industrial revolution but a new world order as well. The ideological divide between liberalism and authoritarianism is remerging and is finding a new battleground in the digital field. The Internet, the world's most important medium for business, trade, and information, is splitting into camps similar to those of the Cold War. On the authoritarian side of this new "Digital Iron Curtain" are China and Russia, where citizen data are centralized and AI algorithms are trained to restrict citizen liberties and freedoms, as well as nudge them into specific, state-friendly behaviors and thoughts. On the opposing side, the liberal camp is currently divided, as the United States and the European Union seem to have different perspectives on how to deal with citizen data.

Relatively unrestricted access to citizen data by corporations in the United States has led many to fear the rise of private surveillance.[9] Shoshana Zuboff, in her book *The Age of Surveillance Capitalism*, argues that unimpeded and unregulated harvesting and analysis of citizen data by big corporations has created new markets that are highly exploitative and insidiously undermine the foundations of liberal democracy. She gives the example of the apparently benign game Pokémon Go. While users think they hunt Pokémon around a city, the creators of the game are using their data in order to direct them to the front doors of McDonalds, Starbucks, and others. The Pokémon hunters are thus being sold without their knowledge. Such exploitation of user data is at the heart of how the business models of digital giants Google and Facebook work. Zuboff claims that every time we search for something or connect with our friends on social media, we are being analyzed and nudged to behave in one way or another, while being constantly watched, indexed, clustered, and correlated—and ultimately sold to the highest bidder. This insatiable quest for aggregating user data is bound to amplify many times over in the near future, and interconnected consumer products in the Internet of Things could become surveillance objects. Surveillance capitalism is therefore almost identical to the digital authoritarianism of China, with the only difference being that, in the former, power is exercised by private corporations instead of governments. This emergent, authoritarian, "digital corporatocracy" in the West—and more specifically, in the United States—is overpowering potential competitors,[10] colonizes other industries,[11] and lobbies for more concessions. The most important concession of all is, off course, further centralization and unhindered access to citizen data.

The political landscape in the European Union is considerably different than that of the United States. There, regulation is restricting access to citizen data and tries to create institutional defenses against the march of the Silicon Valley giants. The two most important pieces of European legislation are the European Union's GDPR, implemented in 2018, and the EU Copyright Directive that since 2019 has started to come into force across EU countries. Although built on good intentions, both those pieces of legislation are problematic. For example, the GDPR calls for the "right to be forgotten" in search results by claiming that this right stems naturally from the right to privacy. This provision has led to controversial court rulings, such as the one in 2019 whereby Finland's highest court ruled that a convicted murderer had the right to have public information about his crime removed from Google search results.[12] Moreover, the GDPR did not anticipate the emergence of distributed ledger technologies that, as we shall see in detail in a later chapter, do not allow the erasing of past data but, nevertheless, provide extremely potent solutions to democratizing the digital economy and solving for the problem of privacy too. Both pieces of European legislation increase compliance costs without discriminating between large and small business—and thus favor the former to the detriment of innovation by less capitalized start-ups. More specifically, the EU Copyright Directive, designed to target big social media platforms, is imposing near impossible compliance demands—for example, by enforcing EU regulation in threads of content submitted by non-EU citizens residing in countries that may have different laws concerning fair use of expiring copyrights. The arbitrary imposition of astronomical fines by the European Union means that the EU Copyright Directive is likely to result in the walling off of the content created in Europe by the rest of the world. Because of such measures by liberal regulators, and in combination of digital authoritarian states raising content firewalls to control their citizens, the Internet is fragmenting. It is not a "global village" anymore, but a collection of continents and islands, each with different cultures and norms, drifting apart.

Meanwhile, and despite the shortcomings of how liberal governments on both sides of the Atlantic try to deal with the rise of surveillance capitalism, there are clear, and hopeful, signs of an emerging awareness that digital giants must not be left unchecked. In 2019 the Federal Trade Commission levied a US$5 billion fine on Facebook for misusing user data in several cases, including the case of Cambridge Analytica.[13] It was the largest civil penalty the agency had ever handed out. In the same year the European Union fined Google US$1.7 billion for advertising violations, which brought the total amount of fines to Google by the European Union since 2017 to US$9.3 billion.[14] Increasingly louder voices suggest that the oligopolistic practices of big tech must be curbed either by setting limits or by breaking them up.[15] They cite precedents such as AT&T's breakup in the 1980s and the limits put on Microsoft during the Internet's rise in the 1990s.

However, all such measures may still prove to be inadequate. They may have worked in the past, but this time things are rather different. The debate around data privacy and digital oligopolies is taking place as liberal democracies are being challenged externally by an increasingly self-confident China, and internally by a multitude of illiberal populists. Our system of government is finding itself at a tipping point, and how we deal with the unique challenges of the new, fragmented, AI-powered digital economy may decide its future survival. Sharing prosperity in the Fourth Industrial Revolution requires a new playbook and a new economic model for capitalism. Breaking up Facebook and Google, or setting limits on how much personal data companies can collect, may end up as mere cosmetic interventions with little long-term effect. We need something far more radical. Given the exploitative and rent-extraction business models of digital platforms, the AI bounty of the future is likely to fall mostly into the hands of the few while leaving the rest of us to share scraps. Unless something changes drastically, our future could be one of citizen indenture to digital corporatocrats that would resemble medieval feudalism in all but name. Given this dystopian double whammy of corporate surveillance and automation-induced unemployment, there are resurgent arguments in favor of nationalizing key sectors of the economy and increasing the role of the state, particularly in providing welfare for all.

THE CASE FOR BIG(GER) GOVERNMENT

In the United States, young Americans and supporters of the Democratic Party are now more supportive of socialism than capitalism by 6% and 10% margins, respectively.[16] Alexandria Ocasio-Cortez, a 30-year old Democratic member of the House of Representatives, is speaking the mind of many of her generation by boldly proclaiming herself to be a "socialist," while her more senior colleague and US presidential candidate Senator Bernie Sanders also favors a more socialist America. In Europe, leftist populists, like Syriza in Greece and Podemos in Spain, fully identify with socialism, while the ex-leader of the United Kingdom's Labour Party, Jeremy Corbyn, championed the nationalization of large parts of the country's economy.[17] In Germany, a country that thirty years ago celebrated the end of socialist East Germany and its reunification under a liberal constitution, another 30-year old politician, Kevin Kühnert, the talented leader of the Social Democratic Party's youth organization, is also evangelizing the return of socialism.[18] There are numerous other politicians, young and old, in many other liberal democracies that are highly critical of free markets and propose that the state should take over the means of production.[19] The COVID-19 pandemic offered a fresh opportunity to such voices to argue that liberal democracies must socialize their economies to a high degree to become both more resilient and fair. Although the range and intensity of "socialist" policies advocated may vary considerably, it would be safe to say that most pundits are in favor of a larger role for government, and they support stricter regulations and a gradual expansion of public ownership over major industries, as well as the extension of the welfare state.

This is not the first time when the ideas of free markets and economic liberty are attacked on the pretext that capitalism has failed and that governments should take over to reestablish "fairness." In fact, there are striking parallels between our times and the early 1930s when the rise of fascism and Nazism was explained as the dying gasp of a failed capitalist system. Like then, we too today tend to view the mostly right-wing illiberal populists as the recurring symptom of that same malaise. We also tend to forget how similar the aspirations and ideologies of those right-wing populists are to those on the opposite side of the political spectrum when it comes to empowering the role of the state while undermining the power of the individual. As the Austrian-British economist and philosopher Friedrich Hayek warned in his book *The Road to Serfdom*,[20] aiming to ensure prosperity by centralizing power around a state inevitably leads to undemocratic outcomes, because it requires that the "will of a small minority be imposed upon the people."[21] Fascists, Nazis, and communists were, from his perspective, exactly the same. Although Hayek did not oppose the provision of welfare by the state, he maintained that the abandonment of individualism in favor of an all-powerful state inevitably leads to tyranny. Hayek was defending classical liberalism in the 1930s, but all those who believe in liberal values should do the same today too. We need to tread very carefully between hedging for a future where most jobs will cease to exist, while retaining and upholding liberal values and not giving in to the sirens of more state control and centralization.

In fact, the state is already dominating most national economies in the liberal world. Capitalism may be in operation at a global scale—because of the globalization of financial markets and international trade—but if one zooms into national economies the story appears quite different. The role and size of the state in most liberal democracies has seen a phenomenal increase over the past decades. For example, in the United States—a country that for many represents the freest market economy in the world—the role of the state has quadrupled over the past century, from less than 7% in 1929 to around 42% of GDP today.[22] The picture is very similar across the European Union, where average government spending is 47.3% of GDP.[23] Citizens are becoming ever more dependent on government support, especially as suppressed incomes and piling debt push many below the poverty line. By way of illustration, since 1989 there has been a 10% increase in the US population, but a whopping 60% increase in the number of people in federal programs.[24] Nevertheless, and despite increased government spending on welfare, poverty and inequality persist. Moreover, increasing the role of government further is making the principal-agent problem worse and goes against the values of classical liberalism, which this book defends, by undermining personal freedom and economic liberty. And yet, we must somehow find a way to deal with rising wealth and income inequality as the tsunami of the Fourth Industrial Revolution starts to hit, and as we internalize the fallout from the COVID-19 pandemic.

UNIVERSAL BASIC INCOME

Faced with the devastating impact of automation on work, advocates of enhancing government intervention policies support the expansion of the welfare state in the form of universal basic income (UBI).[25] UBI proposes that all adults should receive the same minimum income from the government irrespective of their other incomes. This is an idea first proposed by Thomas Paine in the late eighteenth century, when he suggested that big landowners should be taxed and the dividends redistributed to every young man in America. Paine's economic argument rested on the idea of "rent": the owners of producing assets gain from the work of those who use these assets to produce goods and services. This gain is a rent because the owners do not participate in production. In a digital corporatocracy the owners of technology are drawing rents from citizens who are the users and data providers on their digital platforms, just like the feudal lords of medieval Europe or Paine's eighteenth-century America. Advocates of UBI argue that when work will be intermittent or absent because of the domination of AI-powered digital platforms in the economy, UBI could provide a safety net and replace the current, complex welfare systems. Several schemes of UBI have been piloted around the world, in places like Finland, the Netherlands, Kenya, and Canada, with mixed results.

For example, between 1974 and 1978, in a UBI trial called MINCOME, the Canadian government paid the poorest residents of the town Mincome in Manitoba up to CAN$19,500 per year with no constraints on how the money should be spent. Researchers who were able to closely study the effects of this trial reported many positives, such as teenage children completing an extra year of schooling compared to other teens in similar small Manitoba towns and a decrease of 8.5% in hospitalizations with the largest drop in admissions for accidents, injuries, and mental health diagnoses, while the employment rates stayed the same throughout the trial.[26] That trial seemed to illustrate that UBI was a better way to alleviate poverty when compared to a traditional unemployment benefit approach.

On the basis of such a hopeful outcome, the Finnish government launched a UBI experiment in 2016 by giving US$658 (€560) per month to a group of 2,000 adults around the country who would otherwise receive unemployment benefits.[27] The extra income was not taxed at the rate of normal unemployment benefits, and participants were not obliged to seek employment during the two-year trial, although UBI would be lost if they found work. Support for the experiment waned a year later, for several reasons. The total cost of €20 million was deemed too high, while the monthly payments were hardly enough to cover an adult's basic living expenses. From a scientific perspective, there was no control group to benchmark the results of the experiment. Moreover, it was felt that the time horizon of the experiment was too short for people to really change their behavior. So, in late April 2018, the Finish parliament decided to terminate the experiment.

UBI is problematic in several ways. First, in terms of effectiveness, the levels of income support are very low to sustain any meaningful existence other than basic sustenance. Such an effect may only be significant in cases where poverty is extreme—for instance, in a Kenyan experiment where allowing for a minimal income to cover basic needs could really make the difference, as people would be more likely to take risks with entrepreneurial activities that would, in turn, pull them out of the poverty hole. The need for a limited version of UBI became apparent during the COVID-19 pandemic—another extreme situation—in order to sustain the income of millions of displaced and furloughed workers over a relatively short time period. But imagining that one could extend UBI to cover the needs of most people in a Western society undergoing massive transformation due to work automation is probably an illusion. For a start, it begs the question of how UBI would be financed by any government when the working middle class that provides most of the tax income will have become impoverished because of job losses. Supporters of UBI should also take note of the current difficulty that governments have in trying to net taxes from global corporations with sophisticated accounting methods and access to tax-friendly jurisdictions. Moreover, there is a limit to the monetary financing of government debt by central banks. But a question not frequently asked is this: How truly "revolutionary" is the idea of UBI?

As discussed, governments in liberal democracies already redistribute wealth on a massive scale. In the United Kingdom, for example, more than 60% of households receive some state benefit. In 2017 the United Kingdom spent £600 billion in tax, which is around £10,000 per citizen, or approximately £40,000 per family.[28] Given that the average household income is around £28,000, the UK government spent far above that level on a family's behalf.[29] In effect, a "universal basic income" already exists, but the difference is that citizens do not get to see the check.[30] UBI is therefore not something new and radical, but a mostly repackaged "product" that follows the existing trend of making citizens more dependent on government spending. The exclamations of fervent support that UBI receives from many top executives in big, high-tech companies should also give us pause.[31] UBI may appear to "compensate" for job losses due to automation but in reality allows big tech to continue with business as usual and forge ahead with work automation while passing the economic externalities to governments and society.

THE CYBER REPUBLIC PROPOSITION

A more radical approach and a new playbook are necessary in order for liberal democracies to survive and thrive in the Fourth Industrial Revolution, as well as overcome the new challenges of defending liberty and job security in a post-pandemic world. We need to explore options other than simply giving an even bigger role to government and extending the already rather sizable welfare state. As we search for this new playbook, we need to keep individual freedom and liberty firmly at the core of our thinking. With that in mind, this book's proposition for the future of liberal democracy is a "Cyber Republic" based on three core principles: rethinking business models for an AI economy, repositioning AI as a human-centric technology, and extending citizen rights.

A digital economy powered by algorithms and data allows new opportunities to reform capitalism so that it works for the many and not just the few. Artificial intelligence systems need a variety of elements to function, some of which are privately owned and some more diffused and dispersed, such as open-source software. Most notably, those systems need user—or citizen—data. And yet, current digital business models do not reward citizens for the use of their data. But, maybe, "reward" is not enough to solve the wealth asymmetry problem of the Fourth Industrial Revolution. Maybe we need to imagine some form of *co-ownership* and *cogovernance* of the future digital platforms, where entrepreneurs, creators, contributors, investors, and users all have a stake in the economic success of a business venture. Such a business model would demand that the interests of all those different stakeholders be aligned. *Cyber Republic* will examine the potential of distributed ledger technologies in providing solutions both for the democratic governance of digital platforms and for more diffused ownership based on individual contribution.

Beyond rethinking ownership and governance of digital business, we must also rethink how we reposition AI as a human-centric technology, in order to solve for the problems of surveillance, loss of freedom, and, indeed, safety from fully autonomous AI systems going rogue. Zuboff, in her book *Surveillance Capitalism*,[32] correctly points out that anchoring the debate on surveillance capitalism purely on "privacy" is not powerful enough to effect any significant change. Most users are already aware of the fact that they are relinquishing their privacy when they sign up to a platform, or when they access a website and accept cookies. Even assuming that alternative options were available—which are not—the cost of allowing access to a sliver of your data is insignificant in comparison to the benefit of accessing information or a service. Zuboff suggests that we should recast the conversation around privacy as one over "decision rights": the agency we can actively assert over our own futures, which is fundamentally usurped by predictive, data-driven systems. These decision rights can be individual or, in their more powerful version, collective. Arguably, the most effective way for these decision rights to exist would be for citizens to have property rights over their data. *Cyber Republic* will examine how to use distributed ledger technologies to deliver a new construct for managing the value of user data called a "data trust" and how data trusts can become instrumental in democratizing the future AI economy by delivering democratic data governance and fiduciary data property rights. Moreover, *Cyber Republic* will propose how we can re-embed AI in human systems using principles from cybernetics. As we stand today, AI is a threat to our liberal system of government; in a

Cyber Republic the human-centric AI proposed in this book will enhance individual freedom and liberty and enable new forms of advancing individual learning and human potential. But the most radical, perhaps, proposition of *Cyber Republic* is the extension of citizen rights.

EXTENDING CITIZEN RIGHTS

If liberal democracies are to survive in the twenty-first century, they must find a way to solve the principal-agent problem and reestablish citizen trust in liberal institutions. I will argue that for this to happen, the many must be given more power and a more direct participation in political decision-making. But how realistic of a prospect is that? Liberal democracy abhors the direct involvement of citizens in government. As discussed, it was founded with the sole purpose of preventing such involvement from happening. Motivated by the same horror that Plato felt during the Peloponnesian War, the founders of liberalism in the eighteenth century saw direct democracy as the road to tyranny and mob rule. The French Revolution confirmed their worst nightmares: the rebellious public was prone to violence and wild emotional swings and would easily fall prey to populist demagogues. The philosophical hostility of liberalism toward direct democratic rule persisted in the twentieth century, becoming the dominant political dogma in liberal democracies.

Resistance to granting citizens the right to have a more direct say in legislation and policy making rests mostly on economic analysis and the psychology of voting. For example, economist Bryan Caplan has argued in his book *The Rational Voter*[33] that, given the statistical insignificance of an individual vote, it is virtually costless for citizens to express at the ballot box not a rational choice but their misconceptions, prejudices, and tribal loyalties. Caplan's view seems to be confirmed by findings from behavioral psychology. Psychologist and Nobel laureate Daniel Kahneman, in his book *Thinking, Fast and Slow*, writes that "when faced with a difficult question, we often answer an easier one instead usually without noticing the substitution."[34] Using Kahneman's finding, *Financial Times* columnist Tim Harford noted[35] in reference to the EU referendum in the United Kingdom that "the difficult question in the referendum might be, 'Should the UK remain in the EU?'; the easier substitution is, 'Do I like the way this country is going?'" The first—the actual question—required deep expertise in order to be answered. One needed to have knowledge of economic analysis, of the legal obligations stemming from the European Union treaties, of geopolitics to assess risks, and so forth. That is why, Harford claims, voters answered the second, easier question that did not require expertise, a question that anyone could answer on the basis of his or her subjective experience with a binary "yes" or "no."

Referenda are thus considered a bad choice for liberal democracies because of knowledge asymmetries—most voters are simply not knowledgeable enough to decide complex issues.[36] Nevertheless, research shows that citizens in liberal democracies think differently. A 2017 Pew Research Center survey found that 66% of their survey respondents across 38 countries supported direct democracy.[37] This is a strong signal that something is changing in the political consciousness of citizens across the world. Despite what economists and psychologists may say, many citizens nowadays *want to be directly involved* in how their countries are run. The citizen assemblies that spontaneously sprang up in the streets of Athens and Barcelona during the economic crisis are evidence that citizens possess enough wisdom, responsibility, and will to take control of their destiny in a democratic way. But how can we balance the risks of direct democracy with the liberal checks and balances of constitutional law and representational government?

Switzerland is a country well ahead in blending representational and direct democracy. Swiss citizens are empowered to propose national referenda, which are held four times a year. Power is shared among a weak federal government, a lower and upper house of parliament with a limited range of legislative categories, and the cantons and communes that are run by communal assemblies. Switzerland is, of course, a special case, given its history and traditions. Nevertheless, its model of politics can be an inspiration for how liberal democracy could evolve in other countries too. Since all forms of direct democracy result in referenda, we need to address the problems of having large numbers of people voting on issues that may be complex and require a certain level of expertise that the many may not have—the knowledge asymmetry problem. For an assembly of citizens to make rational decisions on complex issues of government there is a need for expert knowledge, as well as trust in expert knowledge. Citizens must acquire a good enough understanding of the issue they are called upon to decide and have some "skin in the game" too. In other words, their decisions must carry some degree of personal responsibility. If we do not find good solutions for the knowledge asymmetry problem, we risk majoritarian rule, which is not democracy but tyranny. Solutions to the knowledge asymmetry problem need mechanisms to ensure that all citizens can access the expert knowledge that is required for them to make decisions on any given matter, to an adequate degree, however complex that knowledge might be. This knowledge may come from various sources, including digital databases, AI systems, and direct communication with human experts.

There are methods from deliberative democracy that we can use in order to deliver such mechanisms. These methods enable voters to be educated about the issue that they must decide upon and get the opportunity to discuss the issue at length before reaching a commonsense consensus. Citizens must also be motivated enough so that they spend the necessary time in the civic duty of participating in democratic decisions. The obvious, technical challenge is how to blend deliberative democracy with representational democracy and, indeed, how to deliver the necessary citizen "education" that solves the knowledge asymmetry problem in a digital sphere crowded by fake news and echo chambers, as well as increasing popular mistrust toward experts. These challenges will be addressed in the rest of the book, and practical solutions that could scale in countries much larger than Switzerland will be proposed.

Finding practical and workable solutions for the knowledge and wealth asymmetries would allow us to imagine an evolved form of liberal democracy for the twenty-first century where the enormous potential and talent of ordinary citizens is unlocked in order to advance human knowledge and well-being, a Cyber Republic. The prize for such transformation of our political system would be enormous. In the ideological battles of the twenty-first century this evolved form of liberal democracy would reignite the flame of hope for the oppressed and provide the intellectual counterbalance to authoritarianism. Without it, our politics will regress toward tyranny of one form or another. So, let's see how we can make a Cyber Republic happen.

6

CITIZEN ASSEMBLIES

The day's persistent drizzle and a murky, overcast sky made Brussels look dull and gloomy, which was not unusual for the middle of January. But, for the two hundred or so people assembled in a large hotel ballroom, not far from the iconic Berlaymont building, the weather conditions did not matter. Sitting around tables in teams of fifteen, they pored over papers, feverishly debating the future of brain sciences. And yet, this was not a conference of brain scientists. Participants were construction workers, students, housewives, office assistants, clerks, accountants, nurses, police officers, teachers, pensioners—anything but experts in the inner workings of the human brain. A retired baker from Antwerp could be seen debating with a young mechanic from Manchester on how the European Union should prevent the medicalization of human cognitive diversity. Next to them, a mother of three and practicing lawyer from Berlin would nod as a young factory worker from Valencia pontificated on incentivizing pharmaceutical companies to invest in alternative treatments. Nine different languages were spoken in the ballroom. To prevent this highly diverse assembly from degenerating into a biblical Babel, interpreters were placed next to the participants in order to enable direct conversations. A team of facilitators kept the whole process in check, while another team of writers and editors collected participants' suggestions and composed them into proposals that were to be put later to a vote. This unexpected assembly of ordinary folks, from all walks of life, from nine different countries of the European Union, had been given the task of providing policy recommendations to the European Parliament for funding research in neuroscience. It was the first political experiment to include nonexpert citizens in the governance of scientific research on such a grand scale.[1] Aptly, the project was called Meeting of Minds.

DEMOCRACY AS A STATE OF MIND

I was one of the facilitators in the Meeting of Minds project. It was a role that gave me a unique viewpoint from which I was able to witness a phenomenon that I had never thought possible: the transformation of a group of unrelated individuals—a ragtag group of ordinary folks that shared little in common other than their ignorance about the science of the human brain—into a formidable democratic polity. It was an experience that changed me in a profound way and made me revise my previous ideas about democratic governance. Until then I was very much a *noocrat*,[2] believing that for a society to be wisely governed the scales of decision-making should tip in favor of expert opinion. Agreeing with Platonic, Hobbesian, and liberal critiques of direct democracy, I believed that only the most intelligent, moral, and well-educated citizens—the "best"—should be allowed to make laws and govern, while ordinary people should just keep quiet, watch, and occasionally vote. Indeed, as the world was becoming more complex and interconnected, I was convinced that only specialized and highly knowledgeable experts could achieve the level of understanding that was necessary to arrive at the right decisions on policy. After all, knowledge was what separated the best from the rest. Therefore, the general public would be better served if allowed the freedom to pursue the fulfillment of their individual lives and interests, while leaving the complicated business of making laws and running a country to the professional class of elected politicians and their expert advisors.

Clearly, such a relationship between the electorate and the elected was not an easy one. As discussed earlier, political scientists refer to the problem of representation as the principal-agent problem. And yet, I was convinced, that this was a necessary price to pay due to the unavoidable knowledge asymmetries in any civilized society: the "agents" have privileged access to expert knowledge that allowed them to make better decisions than the ignorant "principals." Given the vast range of issues that a democratic polity must grapple with, the need for in-depth knowledge in order to comprehend those issues before making decisions, and the time one needs to dedicate in this whole process, the problem of knowledge asymmetry seemed much harder to solve than the principal-agent problem. After all, the social contract in a representative democracy allows for free elections in which the principals can replace their agents and thus hold them to account. Elections enforce some degree of alignment of goals between voters and their representatives, which, as far as I was concerned, was good enough. But my experience as a facilitator in the Meeting of Minds project made me realize that *it was possible to solve knowledge asymmetry* for the principals too. It was all a matter of good design and proper technology. On that gray and wet January day in 2006 in Brussels I underwent my Pauline conversion from a "noocrat" to a "democrat."

The Meeting of Minds project had started a year earlier with the recruiting of citizens from nine EU countries.[3] There would be fourteen citizens from each country, selected at random, but with the requirement that they broadly reflected their country's demographics in terms of age, education, and gender balance. An additional set of requirements included that they should not be brain science professionals[4] nor have any vested interest in the subject—for instance, by being themselves patients with some kind of brain disease, or having a close relative who was, or being members of a brain patient group or association.[5] I remember the first day that I met the first group of citizens. When they were told what they would be discussing, their first reaction was disbelief. "*What do we know about brain science?*" asked one of the delegates. "*How can we possibly advise on something of which we know nothing about?*" I too was deeply skeptical, but, being a facilitator, I made sure to hide my doubts. So I explained how it was important that they did not have any knowledge, or any vested interest, as this allowed them to offer a valuable perspective on the social impact of brain science, something that expert scientists were too biased to offer. But, as I was explaining more of the details of the project, their disbelief quickly turned into suspicion. No one had ever asked their opinion on such complex matters before, they exclaimed. In fact, no one had ever asked their opinion about anything, other than voting for a political party every four years or so, something that most of them thought pointless anyway. So, why now—and why them? Why did their opinion suddenly matter? Was it, perhaps, because the European Commission was colluding with the big pharmaceutical companies in order to legitimize some dark machinations against common folk—whatever those machinations might be? Surely, they were not going to become pawns in that game of corporate greed and trickery! It took a lot of convincing, and some pleading, to get them to at least sign up and explore a little more of the project before they made up their minds. In the end, the citizens agreed to participate on condition that they could quit anytime, should they detect that they were being used and not really heard. And so, with a mix of hesitation, mistrust, and curiosity, the project got underway.

In the first few weeks, each national team got to explore and learn, as much as possible, about the current state of research in the human brain. This learning took place in a variety of ways, but mostly through facilitated dialogue sessions with medical experts, leaders of patient groups, and science researchers. What was significant about this start-up stage was that the citizens realized the value of their role. They did not have to become brain experts in order to debate how brain science might affect ordinary lives. Their role was to provide a *moral direction* for brain science: to come up with ideas on what would be *right for society*—and, therefore, what aspects of brain research should be abandoned, should be modified, should remain as they are, or should be further regulated. As nonexpert citizens, they were perfectly qualified to provide such direction—in fact, they were the *only qualified people* capable of doing so, for *they were* the society, the demos, the end users and principal stakeholders of scientific research. As soon as they fully realized their role, their self-confidence began to grow with every meeting. When the first European Convention took place a few months later, all citizens from all nine nations came to Brussels with a completely different mindset that the one they had had at the beginning: *they now felt equal to the experts.*

The Meeting of Minds project included a start-up phase, followed by the First European Convention, the return of the citizens to their home

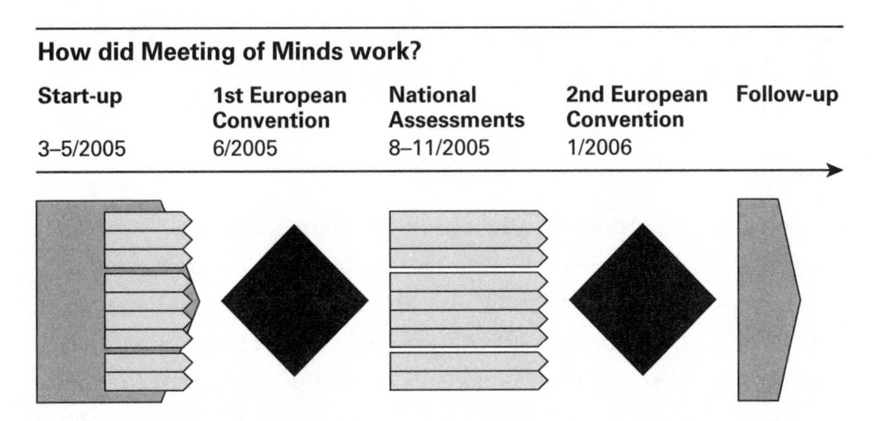

Figure 6.1 The Meeting of Minds took a phased approach with national teams working separately on solving the knowledge asymmetry, then coming together in "European conventions" to debate and decide together with their fellow Europeans. (Image by Meeting of Minds project.)

countries, where they led a number of "national assessments," culminating in the Second European Convention (see figure 6.1). In the First European Convention the debate on brain science was structured around a number of "themes"[6] that the citizens were asked to explore further, with the help of experts. The format used for these deliberations included information systems that relayed what was discussed in the smaller, breakout citizen groups to a facilitating "theme team" that synthesized citizen ideas into proposals that the plenary session could vote on, using electronic keypads. Given the level of technology available at the time, language interpretation, as well as the extraction of "common patterns of thought" from the table discussions, was done manually by teams of human interpreters, writers, and editors. The national assessments enabled citizens to gain a deeper understanding of the brain science, inviting experts in the field as well as other stakeholders groups such as patients and patient group representatives, and further developing their initial assessments. The outcome of the national assessments was then used to feed into the Second European Convention. There, following two days of deliberations structured in a similar way as the First European Convention, the citizens decided on a set of recommendations, which was then submitted to the European Parliament on January 23, 2006, for further consideration. The Meeting of Minds was perhaps the most successful experiment in the application of deliberative citizen democracy for science governance at an international scale.

THE EMERGENCE OF A DEMOCRATIC ETHOS

The quality and breadth of the citizens' recommendations to the European Parliament was not the only remarkable outcome from the Meeting of Minds project. Something else, and far more profound, happened too. As the citizens underwent this process of deliberation, they started shedding their initial reservations and suspicions and began to behave in a very interesting—and for me very unexpected—way. They stopped thinking about their individual interests and started thinking and behaving like "true citizens" performing a civic duty. To understand the difference in thinking, we need to go back to the ancients. Aristotle had made an important observation in his *Politics* that the person who does not care for the good of society (of the "polis") is a liability to democracy.[7] A true citizen is therefore someone who thinks of the common good and *makes decisions together with other citizens for the common good*. In this sense, to become a citizen in a democracy, we must extend our consciousness beyond what gives us individual satisfaction or what improves our individual lives and think of our fellow citizens too, who may be different from us and may have opinions that are in conflict with our own. Democracy demands that we find a way through dialogue to bridge our differences and achieve the essential state for human collaboration, which is *consensus*. We cannot achieve democratic consensus without changing our emotional state first.

What was revealing with respect to the participants in the Meeting of Minds project was that this "civic consciousness" emerged *spontaneously*: not as the result of personal reflection or effort, and certainly not by coercion, force, or manipulation, but as *the natural outcome of people coming together in a meaningful dialogue*. And by "meaningful dialogue" I mean a dialogue that was guided by two critical preconditions. First, *equality*: every citizen had an equal standing and exactly the same rights as everyone else. Second, *dignity*: every citizen was respected, while he or she had the obligation to respect everyone else. This respect was grounded on the intrinsic value of human life and experience and not on academic or professional credentials, wealth, intelligence, age, sex, race, or physical attractiveness. Equality and dignity were the driving forces for the emergence of citizen consciousness among the participants in the Meeting of Minds project.[8] This consciousness was then applied in various practical ways. Citizens started working as well-drilled sports teams; extroverts were reigned in, and introverts were encouraged to voice their opinion. A sense of belonging to a wider community—indeed a truly democratic polity—prevailed. As one delegate declared, "This is the first time that Europe has asked my opinion. I now feel like a true European."

Significantly, civic consciousness held all notions and proposals in check, and extreme opinions and absolutist views were resisted. The citizen assembly of the Meeting of Minds project naturally gravitated to the center and to commonsense decisions where consensus was indeed possible. For example, there was a young gentleman in the Second European Conference who held a very strong view in favor of alternative medicine. His conviction was that modern science and biomedical technologies were pushing people away from their "natural state" and that treating disease with traditional herbs and "spiritual" methods ought to be legislated as the preferred option. Moreover, he claimed that this whole citizen experiment was a farce, and that the European Commission was using the participants in order to legitimize the stranglehold of pharmaceutical companies on public finances and health systems. Surely, this was a conspiracy that citizens had an obligation to reject! The citizen assembly listened to his pitch and voted him down by an overwhelming majority. It was another moment of revelation to me, for not only was he an eloquent and charming orator, but his position was in fact very popular at the beginning of the experiment. So what had caused the shift in the citizens' opinion from the extreme to the center?

Arguably this shift was the result of the citizens having the opportunity to explore the science and gain a deeper understanding of the problem that they were called on to deliberate about. *Knowledge asymmetry was thus reduced.* The shift also correlated with an increased sense of personal responsibility; citizens felt that their views and actions *had consequences*—that, unlike their highly diluted vote for a political party in parliamentary elections, having an informed opinion and shaping a consensus on brain science *really mattered.* Feeling personally responsible for the final outcome motivated citizens to seek a deeper understanding of the issues, explore complexity, and reject extremism and oversimplification—thus generating a positive feedback loop between knowledge and informed opinion. The democratic ethos thus became a shared state of mind and a common feeling that governed citizen behavior.

A DESIGN FOR CITIZEN ASSEMBLIES

"Deliberative democracy," the basis of citizen assemblies, although a term originally coined by political scientist Joseph M. Bessette in 1980,[9] is not a modern idea. Evidently, it was the way the ancient Athenian democracy functioned. In the Athenian popular assembly[10] citizens would consult with expert advisors on matters of the state before reaching a decision. Deliberative democracy is also the way that modern representative democracies function at the level of parliaments and governments, where expert advice is sought through a consultation process. This is often termed "elite deliberation." Lay citizens—the nonexperts—may participate in such elite deliberations as a silent audience, or not at all. As I have argued throughout the book, the exclusion of lay citizens from political decision-making is one of the root causes of today's mistrust in liberal democracy. A key problem that we need to solve in order to regain the lost trust is the tension between experts and nonexperts that can lead to the rejection of expert knowledge by the citizenry, especially when the sirens of oversimplification and populism wail ever so loudly, as they clearly do today. The Meeting of Minds project demonstrated how a well-structured and facilitated citizen deliberation process could solve for knowledge asymmetry in a citizen assembly through a meaningful dialogue between experts and nonexperts, while cultivating and enabling the emergence of new, collaborative behaviors that can lead a diverse group of citizens to consensus decisions, however complex the subject. Citizen assemblies are one of the fundamental building blocks of a Cyber Republic, and it is therefore important to summarize their key design elements.

Figure 6.2 shows a possible design for a democratic decision process using the Meeting of Minds approach. A citizen assembly is formed by selecting by sortition[11] a group of citizens whose diversity reflects the

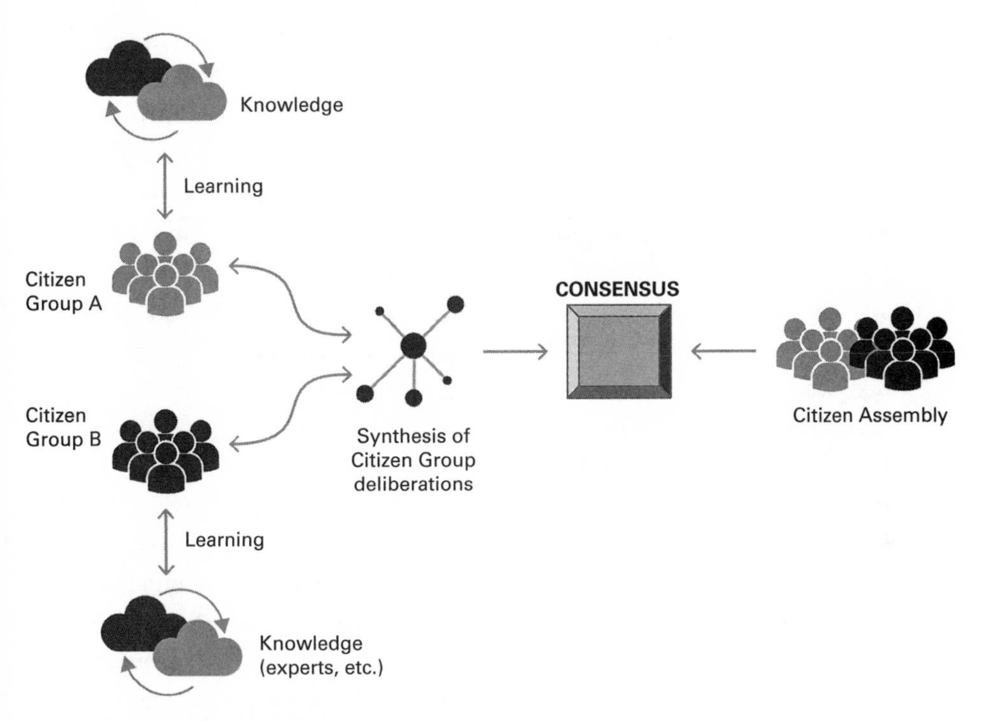

Figure 6.2 Using deliberative democracy to achieve consensus in a citizen assembly. ©2019 by George Zarkadakis.

diversity of the wider social group. The assembly is then segmented into a number of citizen working groups—in the simplified case illustrated in figure 6.2 into "Group A" and in "Group B." The problem examined by the assembly is then broken down in smaller pieces—in this example into two parts. Each citizen group focuses on a different part of the problem and enters into dialogue with a knowledge base, as well as among themselves, in order to learn more. This knowledge base can be a collection of data repositories, libraries, interviews with human experts, digital media, and so forth. Dialogue is an essential ingredient of this iteration, among the citizens in the groups as well as between the citizens and the knowledge base. The result of this iteration is that the knowledge base is also refined, extended, and augmented. The evolved understanding of the deliberating citizen groups is then fed into a mechanism of synthesizing a proposal that each group agrees upon. This mechanism must be able to deal with multiple proposals that the citizens may come up with and facilitate a process whereby the citizens gradually agree on which one proposal is the most acceptable. The Meeting of Minds project did this in a manual way in 2006, but with modern technology this synthesis can be largely automated.

For example, the New York–based technology start-up Remesh[12] is using AI in order to automate the forging of consensus in a group. Cofounders Andrew Konya and Gary Ellis and their team have developed a software application that performs two main services. First, it applies natural language processing (NLP) to pick up semantic correlations in the written text that participants use to express their thinking during a deliberation. The application uses those correlations to identify and select specific participant sentences that capture a spectrum from the "most popular" to the "least popular" ideas, or proposals. Second, the application feeds back to each participant those selected proposals, in an iterative way, in order to ultimately arrive at consensus. The difference between "popularity" and "consensus," and how the two are connected, is illustrated in figure 6.3. Assuming that polling does not take place on the basis of a binary "yes-no" option but allows participants to express degrees of satisfaction for a given proposal, a proposal is considered "popular" when a certain number of group participants strongly agree with it. For example, if we imagine a scale of agreement from 0 (do not agree) to 5 (agree strongly), then the mean of the normal distribution of responses from participants is a measure of "popularity." However, popularity is not enough: think of instances of a vocal minority pushing their agenda through a proposal that others disagree with. In order for any proposal to pass, there must be as wide a consensus as possible; in a normal distribution "consensus" can be measured as the standard deviation, or how wide the bell curve of participants' responses to the proposal is. When there are competing proposals, consensus matters more than popularity. For instance, in the example shown in figure 6.3, Proposal 2 enjoys wider consensus and would be the proposal that passes.

When the phase of proposal iterations is complete, and citizens feel satisfied that they have reached a consensus, groups are dissolved and citizens regroup in the general assembly. Their task is now to discuss and vote on the proposals of the groups. A new dialogue ensues in the

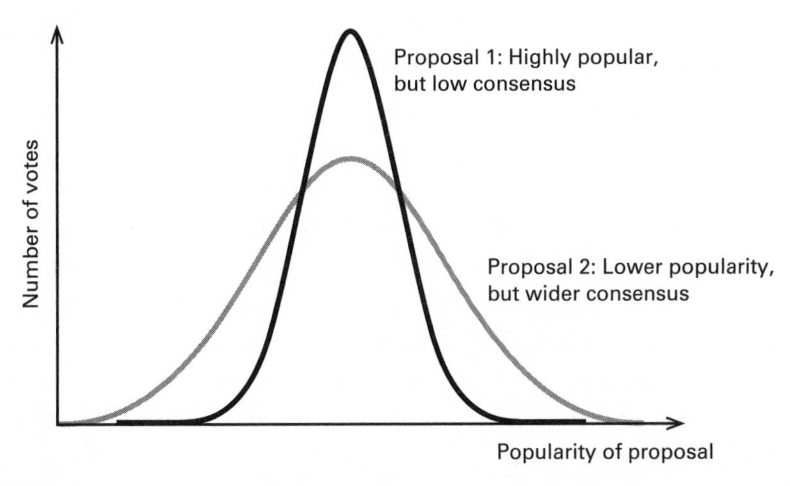

Figure 6.3 Comparing "popularity" and "consensus." The vertical axis represents number of votes, and the horizontal axis the popularity of a proposal. For simplicity, the curves assume a normal distribution of votes (on a range of "agreeing with the proposal" from, say, 0 to 5). Consensus is measured by the standard deviation of the curves (how "wide" the curve is). In the figure, Proposal 1 is more popular but Proposal 2 enjoys wider consensus. ©2019 by George Zarkadakis.

assembly, where representatives of the citizen groups present their proposals and the reasoning behind them. The process that took place during the group deliberations may need to be repeated. Any member of the assembly may also suggest alternative proposals. If consensus is not reached, then the assembly may take a final vote whereby a decision is made by the majority and becomes binding for everyone.

IT'S ALREADY HAPPENING

The Meeting of Minds project used a citizen assembly to explore the ethical implications of neurotechnologies and neuroscience. Citizen assemblies are also being used in many other cases across the world. In her book *The People's Verdict* Claudia Chwalisz looked at around fifty examples of government-commissioned citizens' assemblies over the past decade using this approach to solve policy dilemmas.[13] In Canada, these processes tend to be called "citizens' reference panels"; in Australia, they are often called "citizens' juries"; but they are all based on the idea of including informed citizen voices in policy decision-making. Let's look at a couple of examples.

A constitutional citizen assembly was set up in 2016 in Ireland to deliberate on changing abortion law. Over a period of five weekends, 99 complete strangers, an assemblage of students, housewives, and truck drivers, with views ranging from pro-life to pro-choice to undecided, paved the way to the national referendum.[14] The citizen's deliberations were broadcast on the Internet.[15] They discussed abortion with legal, ethics, and medical experts and listened to testimonies by women who faced crisis pregnancies and had to get an abortion outside Ireland. At the end of the deliberation process the majority of citizens recommended amending the constitution. Indeed, two-thirds of the assembly suggested legalizing abortion without any restrictions. The view and decision of the citizen assembly reflected the national vote in the Irish referendum of 2018. The Irish citizen assembly thus helped to achieve broad national consensus on a highly polarizing subject.[16]

Another example of a citizen assembly used to resolve a major political impasse took place in South Korea. The country is among the top five producers of nuclear energy, but the 2011 Fukushima meltdown in Japan and a 2016 earthquake in Gyeongju, South Korea (a region that hosts six nuclear power plants within 50 kilometers, or about 31 miles, of population centers), polarized the public debate on whether South Korea should continue investing in nuclear energy.[17] The debate was quickly reduced to a clash between economic growth, favored by the nuclear industry and generally older citizens, and safety, favored by environmentalists and the young. In May 2017, the newly elected government of President Moon Jae-in decided against holding a referendum.[18] As discussed, referenda are blunt instruments; citizens do not have the chance to be informed adequately and in sufficient depth about the issue they are deciding upon, and results can lead to deeper social divisions. Instead the Moon government decided to resolve the issue using a "deliberative poll,"[19] which is another way of describing a citizen assembly. Thus, a representative sample of 500 citizens was selected on the basis of administrative district, gender, and age group and was sent briefing materials.

A month later the citizens group was brought together for three days of discussion with neutral moderators and pro- as well as antinuclear experts. Discussions were broadcast throughout the country. The final vote in October 2017 of the citizen representatives was unambiguous: 60% of the respondents voted to resume construction of two nuclear plants at Shin-Kori where more than US$1 billion had already been invested. Yet, 53% voted to decrease the share of nuclear in the country's energy mix. This balanced, commonsense outcome captured the nuances surrounding nuclear energy. A simple yes-no decision—that is, a typical referendum outcome—would not have been appropriate, or as powerful, in resolving the issue. The Moon government accepted the outcome of the deliberative poll and adjusted South Korea's national policy accordingly. Remarkably, following the citizen deliberations and decisions, violent protests against nuclear energy stopped. The citizens of South Korea had reached a consensus.

Examples of citizen assemblies such as the above point to the feasibility of implementing direct citizen participation in complex decision-making where knowledge asymmetries lead to misunderstandings, polarization, and impasses at the level of representatives. They also point to the opportunity for evolving liberal democracies toward a more inclusive model of government. So, how we can make this happen?

ESTABLISHING CITIZEN ASSEMBLIES AS A NEW LIBERAL INSTITUTION

For any degree of participatory democracy to become a realistic proposition, deliberative processes for citizen participation in decision-making *must be credible and scalable*. Credibility is perhaps the harder to achieve. Take, for example, what happened in France following the yellow vest demonstrations in 2018. President Macron, faced with what amounted to a citizen rebellion, launched the "Grand Débat"[20] with the very ambitious goal of engaging with citizens in order to enhance French democracy. Among the various channels[21] that became available to citizens for voicing their concerns were a series of randomly selected citizen assemblies in each of the thirteen French regions and five overseas territories, and another bringing together young people.[22] Most criticism of those citizen assemblies focused on their rushed and flawed implementation.[23] The French government hired too few professional facilitators, making it impossible to guarantee consistency of outcomes across all the assemblies. Participation in the assemblies was not mandatory and thus quickly became self-selecting. This resulted in assemblies becoming gathering places of predominantly well-educated, wealthy, urban, pro-European males whose views were highly unrepresentative of the wider public.[24] Significantly, there were no resources available to educate citizens on the issues in order to solve for knowledge asymmetry. Furthermore, the outcomes of the citizen assembly deliberations, as well as outcomes for all the other channels available to citizens, were nonbinding but were passed to a committee of "wise Europeans" who were tasked with "distilling" citizen proposals into policy recommendations. By doing so, President Macron had profoundly prioritized noocracy over democracy, which made the whole Grand Débat appear farcical. With all those flaws, France's implementation of citizen assemblies did not achieve the level of credibility that was necessary for citizen assemblies to become a new, and trusted, liberal institution. The only tangible outcome was a slight improvement in President Macron's ratings, which evaporated soon afterward.

Successful implementations of citizen assemblies require that the design principles discussed earlier are followed, and that their role and degree of influence are clearly defined from the outset. This role may range from consultative to a national parliament, or to local or national government, all the way to informing the options of a national referendum. As long as the selection of citizens is transparent and random, the goals of the assembly defined, and the deliberations open to scrutiny, other citizens who are not participating in the assembly are likely to agree with the consensus reached.

In comparison to credibility, scalability should be relatively less hard to achieve. Setting up and running citizen assemblies on a regular basis is currently very costly, due to the many manual tasks that need to be carried out, as well as due to high costs relating to coordinating the deliberations of citizens, their dialogue with experts and other stakeholders, and the communication of the process and its outcome to the wider community, a large country, or a federation. The French government tried to scale citizen assemblies at a national level but failed because of the high costs. However, these setup and running costs could be significantly reduced with the better use of technology that incorporates—as we shall see in following chapters—citizen data, AI algorithms, and distributed ledger systems.

But there is another cost that affects scaling, even more prohibitive than the costs of setting up and running a citizen assembly: the *cost of citizen participation.* Participating in politics takes time away from the limited time that we all have to pursue our individual goals. As political scientist Josiah Ober highlights in his book *Demopolis,*[25] each one of us invests our available time in securing two main categories of goods: subsistence (having a job, setting up a business, making investments, etc.) and socially valuable goods (entertainment, holidays, family, friends, etc.). Given that the overall time one has available is inelastic, and that the time we must spend on securing subsistence is inelastic too, participation in political deliberations means that citizens must sacrifice time spent on securing social goods. In an oligarchy such sacrifice is not necessary. Since there is no need for citizens to care about government or politics, the time citizens spend pursuing their own goals can be maximized. Oligarchies, including liberal representational democracies, offer "freedom" from civic duties and enable the full amount of time that a citizen has to be spent on subsistence and socially valuable goods. Nevertheless, if citizen assemblies are to become a new liberal institution and earn legitimacy by truly representing the demos, citizens from all backgrounds must be incentivized to participate.

The Meeting of Minds project solved the problem by covering the expenses of the citizens, as well as providing them with a small stipend for their participation. Without that financial incentive by the project funders, it is doubtful that the citizens would have willingly carved out time from pursuing their socially valuable goods in order to discuss brain science. If citizen assemblies are to become the norm, they must provide some financial incentive to compensate for citizen time. Hopefully, that financial incentive would not need to be considerable, for there are "hidden benefits" in directly participating in politics that are perhaps more valuable than a stipend. As Josiah Ober argues,[26] the only way that citizen participation can take root and scale is when the total value to the individual produced by combining civic duties and social goals exceeds the value produced by pursuing only social goods. Put more simply, citizens who participate in any form of direct democracy should derive significant value from their participation. The Meeting of Minds project demonstrated what that significant value might be: the "*experience of citizenship.*"

The personal transformation that most citizens underwent because of their participation in the project should be considered as evidence of that experience of citizenship. Participatory democracy encourages us to rediscover an essential part of our human nature, the experience of being members of a polity, the feeling of citizenship, and responsibility toward others. We are thus given the opportunity to grow morally, emotionally, and politically and become truly adults, in the full sense of the word, that is, truly *wise* men and women. When participation in political decision-making is grounded in dignity and equality, then the value we derive from this participation vastly exceeds the value we would have derived by dedicating ourselves to the pursuit of our own social goods. The experience of citizenship may explain why the ancient Athenians defended their political system so vehemently, and why the modern Swiss are so proud of their democracy and way of life.

TRUSTING THE DEMOS TO GOVERN

Adopting a measure of direct democracy based on the deliberative model of citizen assemblies, in combination with representational government, could potentially restore citizen trust in democratic institutions and revitalize liberal democracies. The problems of scalability and cost could be addressed using existing technologies that can be easily accessed using a simple smartphone, as Remesh has demonstrated. Incorporating citizen assemblies into the process of government, at the local and the national level, is therefore mostly a political decision. The democratization of our societies has been a gradual and painful process over many centuries. It is now time we took the next evolutionary step by extending citizen rights and including everyone in directly influencing policy making and laws.

We should do so not only because of the obvious threat of democracy disappearing under the multiple stresses of the Fourth Industrial Revolution. An additional motivation for diffusing political power to the many should be the complexity of problems that we are facing in the twenty-first century. To find workable and popular solutions to such complex and interconnected problems—for example, climate change, food insecurity, terrorism, poverty, and the ethics of AI—expert knowledge is not enough. We are witnessing today a systematic denial of scientific knowledge by increasing numbers of citizens. Such extremist opinions are amplified via the megaphones of social media and recommendation algorithms that prey on controversy. Climate change denial, refusal to inoculate children, and conspiracy theories claiming that the earth is flat and that astronauts never landed on the Moon are symptoms of the ever-widening gap between experts and nonexperts. The geopolitical endgame of such civic backlash to scientific and technological progress is that liberal democracies will lose their economic and military advantages.[27] For science and technology to continue on the path of progress, direct citizen participation in political decision-making is therefore absolutely necessary. We need the direct involvement of nonexperts, of lay citizens—the demos—to intervene and provide the moral compass and the necessary common sense, as well as vital societal consensus. As demonstrated by the Meeting of Minds project and the many other successful implementations of citizen assemblies around the world, common sense and a democratic ethos are as much an essential element for legitimizing collective decisions as is expert knowledge.

7

AI FOR GOOD

Debates on AI consistently juxtapose humans against the machines. Artificial intelligence systems will replace us in the workplace, nudge us into obedience, spy on our innermost thoughts, and ultimately overpower us—if one adheres to the dystopian prophecies of the so-called AI singularity. Therefore, AI must be managed and regulated lest humans become collateral damage to technological progress. This spirit of "humans against the machines" has inspired the European Union's GDPR, as well as the many initiatives in AI ethics that seem to be popping up everywhere nowadays. To hedge against the negative consequences of AI, economists and ethicists are trying to find common ground and suggest how to balance the economic potential of AI with its ethical implications.[1] The automation of work by AI is a fine example of this precarious and much-sought balance between opposing needs. Markets tend to ignore the "social value" of work and focus squarely on efficiency and cost reduction. This creates many negative externalities—for instance, a steep increase in cases of depression among the unemployed followed by the breakdown of social cohesion, which are costs that must be borne by society at large. This is not dissimilar to a factory producing excellent goods at low prices while polluting the environment—and not paying for it. So, as we think about the future of AI, there are many who are trying to find ways to repurpose intelligent systems to deliver social goods and not just cost efficiencies. For instance, economists Acemoglu and Restrepo suggest that societies should choose to invest in the "right type" of AI, one that minimizes negative externalities, and deploy it in areas such as education and health care where most people would benefit.[2] They also suggest that tax incentives should be given to companies so that AI systems create new and higher-value work for humans rather than simply automating tasks.

All such proposals are very useful in addressing the impact of AI and its current model of application that benefits mostly the big corporate oligopolies. Nevertheless, these proposals implicitly accept AI technology in the way it is being developed today as a given. In other words, they do not question the foundational nature of AI systems or whether there is an alternative way to design these systems before embedding them in human society. They also largely ignore, or underestimate, a particular characteristic of AI systems that is unique, namely, their potential for autonomy. Autonomy is different from automation. Conventional computer systems automate business processes by encoding in a software program specific steps that the machine has to execute in order to arrive at a logical output. The human programmer determines these steps and is in complete control of the process as well as the outcome of automation. But when it comes to AI systems, the human programmer becomes irrelevant beyond the initial training of the algorithm, and the system evolves its own behavior and inner complexity as it crunches more data. This self-evolution of AI systems is akin to how biological systems learn by interacting with their environment to evolve adaptive behaviors that we perceive as intelligence and autonomy. The highest degree of autonomy is, of course, free will, which we currently defend as uniquely human, although who is to say whether an autonomous AI system won't exhibit "free will" in the not too distant future. As discussed, current research and effort in AGI aim to develop exactly such systems in the next ten years.[3] The increasing autonomy of future AI systems suggests that humans will have even less control over how those systems will behave. Ultimately, a superintelligent autonomous system will be uncontrollable and capable of setting its own goals that may be different than, or in conflict with, human goals.[4] Framing the debate on AI as "us versus them"—as we do today—leads to a future battle that we humans are bound to lose, regardless of how smartly we try to regulate this technology for good. Moreover, the outcome of that battle may pose an existential risk to our species and civilization that goes far beyond the aggravations of economic inequality.[5]

We should therefore reexamine the idea of AI autonomy, understand its origins, and ask if it is a one-way street. Can there be only one future, where intelligent systems and humans struggle to coexist in a tense and uneasy relationship of constant competition and suspicion? Or is there another way to think of machine intelligence, not as something separate from human experience and society, not as the mechanical "other" that is to be feared and regulated, not as the means of reinforcing the dominance of the few over the many, but as something that is purposely built to improve the human condition? To answer these questions, we need to start by tracing the historical roots of AI and identify the moment that gave birth to the idea of machine autonomy. To do so, we need to go back at the dawn of the computer era in the 1950s and 1960s.

CONTROL AND COMMUNICATION IN THE ANIMAL AND THE MACHINE

Artificial intelligence was born out of the dominant scientific and engineering movement of those postwar times called "cybernetics." The goal of cybernetics was to understand how self-organization occurs in complex systems, but also how emergent systemic behaviors affect the parts that make up those systems. A system is generally described as "complex" when its behavior is different from the behavior of its parts. Take, for example, a flock of starlings; each starling has a certain way—or pattern—that it follows when in flight, but the flight pattern of the flock is completely different, as everyone who has had the joy of watching this beautiful phenomenon knows. The flock exhibits an emergent behavior, something that would be impossible to predict if one knew only how any single starling flies. Such emergent phenomena of self-organization occur all around us and at every scale, from colonies of microscopic algae to the neuronal networks in our brains, the physiology of animals and humans, weather systems, biological ecosystems, and, of course, the economy. They can also occur by design in complex engineering systems.

Cybernetics studies the relations between the parts and the whole by analyzing how communication takes place between the parts of the system, looking into what are the various communication pathways that signals have to travel between the parts, what are the messages carried by signals, and how the messages or other characteristics of the signals change over time, and why.[6]

One of the most profound ideas in cybernetics is "feedback." Cybernetic systems exhibit emergent behaviors because the communication signals between the parts are traveling in feedback loops; output signals are often mixed with input signals from the environment or other parts of the system and are fed back into the system and so on. Thus, a cybernetic system operates in a recursive way. There can be numerous feedback loops at play in complex systems. What makes feedback loops significant is that they change the internal structure of the system. Take, for example, the human brain, which is made up of billions of interconnected neurons. In order for the brain to make a decision (conscious or not), it processes input signals from the senses, as well as signals from inside the brain that continuously alter the strength of the interconnections between the neurons. The "brain system" is constantly reconfiguring itself as it interacts with the environment, with itself, and with the rest of the body. Like the brain, all cybernetic systems adapt their behavior to changes in their environment and in themselves through sensing and processing multitudes of signals traveling in feedback loops. Cybernetics' great insight was that this process of adaptation by feedback loops is in effect a process of "learning." In the case of the human brain this is quite obvious: the mechanism of memory is based on how our brain's internal structure modulates over time, and how the interconnections between the billions of neurons are either reinforced or weakened.[7] The strength and pattern of neural interconnections are further controlled by neurochemical reactions that are modulated via feedback loops.[8] The neurobiological mechanism of memory is thus founded in how the internal structure of the brain changes while adapting to both internal and external stimuli and states. However, because of the generality inherent in cybernetic theory, any complex system that adapts its internal structure to a changing environment in order to achieve its goals can also be regarded as a "learning system."

It thus became one of the most audacious ambitions of cybernetics to develop design principles for how humans and machines could communicate, collaborate, and learn from each other. In 1950 the father of cybernetics, Norbert Wiener, published a book entitled *The Human Use of Human Beings*,[9] in which he expanded on the idea that an ideal future society would be one where machines do all the tedious work, thus freeing the humans to perform more sophisticated and creative tasks. When we look at a future where AI and robotics take over human work, Wiener's prophetic vision begins to resonate very strongly. However, there is a fundamental difference between how we debate about AI today and how Wiener and other cyberneticians thought about the role of automation technology. Wiener's concept of human-machine collaboration was based on cybernetic principles of knowledge that required humans firmly embedded within a cybernetic "supersystem." These design principles start with the idea that some measurements (data) are transformed into organized information (a database), which can then be turned into knowledge through correlation, reasoning, and curiosity (a knowledge base), with the end result being some kind of action. In a cybernetic system, sensors, computers, hydraulics, electronics, and humans form part of a supersystem that increases its collective knowledge as it interacts with the environment over time and thus also increases its degree of autonomy. In a cybernetic scenario, machines and humans are coupled inextricably as they both advance together toward a common goal.

It was from within that intellectual environment that the idea of AI was born. But something profound happened at the historical Dartmouth Workshop of 1956 that founded AI. The aspirations of the new science of AI clearly diverged from the original ethos of cybernetics science, as described in the text of the original proposal: "The study is to proceed on the basis of the conjecture that every aspect of learning or any other feature of intelligence can in principle be so precisely described that a machine can be made to simulate it. An attempt will be made to find how to make machines use language, form abstractions and concepts, solve kinds of problems now reserved for humans, and improve themselves."[10]

This foundational statement signified the historical point of departure of AI from cybernetics. The new science of AI was now taking a different view by claiming that it was possible, indeed desirable, to build knowledge machines with complete autonomy. Artificial intelligence thus decoupled humans from machines. This meant that humans were no longer necessary or relevant. Indeed the language of the proposal states that the aim of AI was to make machines at least as intelligent as humans, and probably more so. Artificial intelligence was thus founded as a competition against humans. Progress in the development of powerful digital computers played an important role in AI departing from cybernetics, focusing attention on the machines themselves and how to apply the cybernetic concepts of learning by feedback loops inside a machine.

This core philosophy of AI, of setting intelligent machines and humans apart and often in competition, applies to both its schools of thought, symbolic as well as connectionist. In symbolic AI it led to autonomous systems wherein knowledge was coded in logical rules.[11] In connectionist, probabilistic AI, autonomy is manifested in how deep learning systems operate. For example, in a typical setup for supervised learning, training data are labeled by humans and are fed into an artificial neural network. The human contribution is pure input; there is no feedback loop from the machine back to the human that changes the human. Only the artificial neural network changes its internal structure (i.e., the numerical weights of its nodes) through iterations with human-labeled data. In the case of unsupervised learning or reinforcement learning, the AI system is totally decoupled from humans and learns by itself, as the system AlphaGo Zero by DeepMind demonstrated.[12] The history of AI is marked by "milestones" where machines beat humans: in chess, in *Jeopardy!*, in Go, in cancer diagnosis. Such milestones are celebrated as signs of technological progress wherein the machines improve themselves and become more intelligent than humans in narrow domains until "artificial general intelligence" comes along and machines become smarter than us in everything.[13] As the authors of a paper on AlphaGo Zero proudly stated, "Starting from tabula rasa, our new program AlphaGo Zero achieved superhuman performance, winning 100–0 against the previously published, champion defeating AlphaGo."[14] By celebrating their system's victory over the human, the authors were expressing the dominant AI ideology of the past sixty years, which implies a future in which machines surpass humans. Nevertheless, there is another way to think of AI.

SEARCHING FOR AN ALTERNATIVE AI

Let us return to cybernetics and explore how a different kind of AI may be possible, one in which humans and machines are not adversaries but partners working together toward a more human-centric world. By way of illustration, let us take conversational agents, where current approaches of NLP are used in applications such as chatbots and digital assistants.[15] To reimagine these human-machine conversations using cybernetics, we may look into the work of Gordon Pask (1928–1996), an English cybernetician who made significant contributions in a number of scientific fields including learning theory and educational psychology. Pask developed a theory based on the idea that learning stems from the consensual agreement of interacting actors in a given environment, which he called "conversation theory."[16] Conversation theory postulates that knowledge is constructed as the result of a conversation, and that this knowledge then changes the internal structure of the participants in that conversation. These participants may be humans but may also be humans and machines engaged in dialogue. Pask suggested that the knowledge resulting from such conversations could be represented in what he called "entailment mesh."

Entailment meshes encapsulate the meaning that emerges in a conversation by two participants—say, between A and B. For participant A, the one who initiates the conversation, the mesh encapsulates the topics of the conversation as well as what makes them different, in what Pask termed "descriptive dynamics." For example, say A wanted to talk with B about immigration. The descriptive dynamics of A's entailment mesh would include concepts such as "number of immigrants," "economic impact of immigration on GDP," "multiculturalism," and so forth. She would then need to explain how these topics interact to make a new concept, for example, a "policy for immigration control" (this aspect of the entailment mesh is called "prescriptive dynamics"). Participant B would then need to deconstruct the meaning of A's entailment mesh and reconstruct it in terms that he would understand, then feed this reconstructed understanding back to A. A would then need to change aspects of her entailment mesh, both descriptive and prescriptive, in order to "close the gap" of understanding between her and B. Through such conversational iterations between A and B, a common understanding is achieved, which can be the basis for collaboration, negotiation, agreement, or action.[17]

For comparison, let us look at the most advanced conversational agent we have today that uses the traditional, connectionist approach in AI. In 2019 OpenAI released a limited version of a conversational system called GPT-2.[18] The system was trained on a data set of around 8 million web pages, making it capable of predicting the next word in any 40 GB of text. This relatively simple capability was enough for the system to generate impressively realistic and coherent paragraphs of text, as well as beat many benchmarks in reading comprehension, summarization, and question-and-answer conversations. Nevertheless, there were also certain limitations that are typical of AI systems that learn on preexisting data. For instance, GPT-2 would perform well with concepts that were highly represented in the training data (such as Brexit, Miley Cyrus, Lord of the Rings, etc.) and poorly with unfamiliar texts. Researchers are currently trying to address this weaknesses in the current approach of probabilistic AI, for it fails to capture new meaning in novel conversations or texts, as well as support conversations beyond simple questions about a priori known facts.[19] It is also telling that OpenAI restricted the release of the full training model of GPT-2 based on concerns over the malicious use of the algorithm in deep fakes. In other words, our most advanced conversational systems today are deemed "dangerous" for society because they can hack our brains and restrict our freedom. Arguably, this is happening because current AI aims to imitate humans by adhering to the core ideology of system autonomy. While doing so, AI systems also struggle to become the truly collaborative tools that we actually need.

Pask's entailment meshes could offer an alternative, and more human-centric, route toward artificial conversational agents. Conversation theory principles could be used to connect concepts in the context of a conversation and the meaning that needs to be conveyed. Second, concepts would evolve as a result of the conversation, which means that the memory of the conversation can be implicit in an entailment mesh. This memory may also signify how a particular participant likes to approach learning, what his or her style is. Do participants prefer to be given the "big picture" up front—what Pask termed as "holists"—or do they learn quicker through examples. Entailment meshes can therefore be used to improve personalization. Moreover, this dynamic coevolution of humans and machine—or "knowledge calculus" to use Pask's terminology—is much closer to how our brains learn.

Ultimately, this example of the cybernetic approach of entailment meshes—an instance of "cybernetic AI"—could result in human-machine systems that have goals that are shared between humans and machines. This is in contrast with today's AI that is decoupled from the humans while advancing its own goals. Therefore, to de-risk the existential threat of AI and deliver a human-centric future, it is not enough to simply regulate AI and hope for the best. Instead, we should adopt a top-down approach in how we build intelligent human-machine systems based on cybernetics. This holistic approach requires the design and construction of learning systems wherein humans and intelligent machines collaborate to achieve common goals. These learning systems can be then embedded in any human organization, commercial or noncommercial.

By deriving inspiration from Pask's approach in building conversational machines, we could also make AIs exhibit behaviors that go beyond the mere exchange of a conversation and reach a level that we generally associate with "theory of mind."[20] Theory of mind is one of the most important cognitive abilities of the human brain and is essential for empathy and understanding. We develop this ability between the ages of three and four, when we begin to "hypothesize" that other people around us also have "minds" that are different from but also similar to ours. It is because of theory of mind that we have the ability to empathize with other people's feelings and intentions. Theory of mind is the cognitive mechanism of moral behavior; it makes us feel how others feel—or would feel—as a result of our actions. The memory of conversations, as embedded in entailment meshes, could deliver artificial systems that empathize with humans and help us achieve our personal goals. Current AI approaches lack such cognitive ability, as they personalize through mere statistical segmentation—that is, by grouping us with others with "similar" characteristics. In the context of a democratic polity, and since language is the medium of human communication, politics, and consensus—and theory of mind is the key to empathy—we would need to envision and develop a new breed of conversational agents capable of emulating empathic behaviors. Instead of restricting access to such AI systems for fear of limiting our freedom—as OpenAI did with GPT-2—we should be doing the exact opposite: designing, developing and deploying systems that amplify our potential and provide us with more degrees of freedom and liberty. Let's see an example of how this could be achieved, in the case of citizen assemblies.

EMBEDDING CYBERNETIC CONVERSATIONAL AGENTS IN A CITIZEN ASSEMBLY

We saw how citizen assemblies have the potential to solve knowledge asymmetries between experts and nonexperts by leveraging the dynamics of deliberation and dialogue. Setting up citizen assemblies is currently very costly and cumbersome. Nevertheless, it is not difficult to imagine how we could apply information technologies and develop software applications that can reduce those costs and thus make citizen assemblies easier to set up and, therefore, more frequent and popular. And we could go several steps further than that, by borrowing ideas from Pask's conversation theory, imagining intelligent conversational systems that exhibit empathy and theory of mind and using them to facilitate citizen assemblies.

Figure 7.1 depicts a potential design for a citizen assembly in which cybernetic conversational agents enable facilitation, translation, recording, and reporting. Such agents could free up resources and allow for scalability. In order for those agents to not simply automate but become part of the overall human-machine system of democratic deliberation, they must perform the following tasks:

Understand the context of the debate, that is, what the assembly and the groups aim to achieve. This understanding can be effected through a human-machine dialogue whereby entailment meshes encapsulate the evolving consensus among citizens.

Facilitate knowledge discovery and acquisition; the intelligent agent acquires (or "learns") meta-knowledge about a citizen's goals in order to help the citizen find and acquire necessary knowledge. This is different from an AI search based on keywords. Often the citizen would not know a priori what knowledge is required in order to search for it. Therefore, the agent needs to engage in a conversation with a citizen, understand the citizen's learning style, and use an entailment

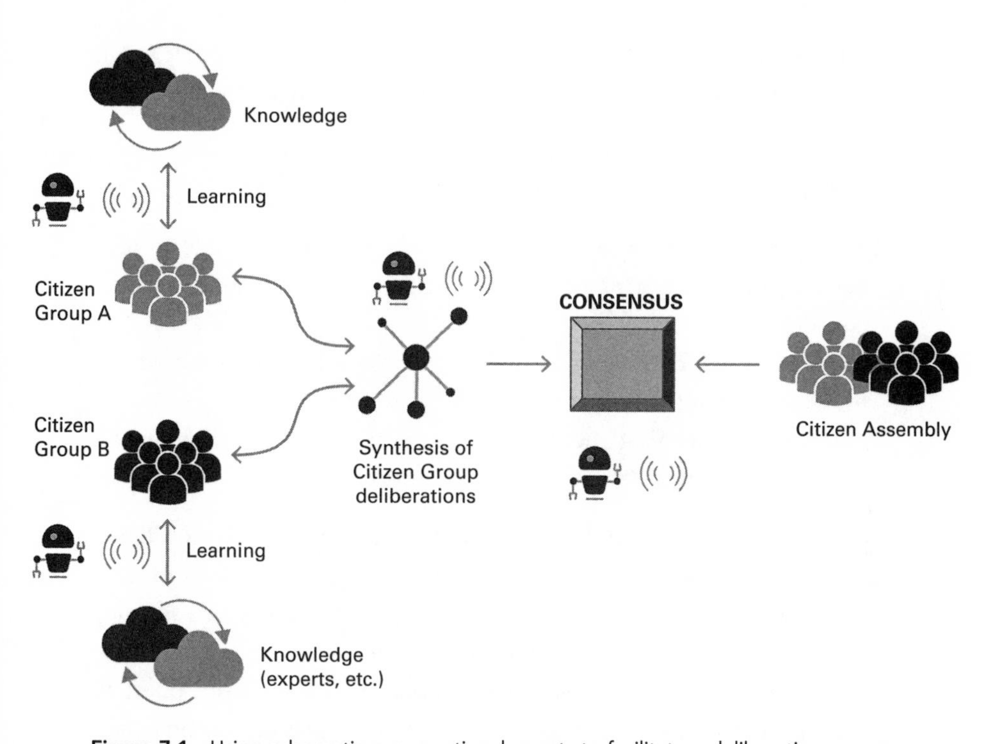

Figure 7.1 Using cybernetic conversational agents to facilitate a deliberative process while preserving humans in the loop. ©2019 by George Zarkadakis.

mesh to advance the combined, and evolving, knowledge of both the agent and the human around a specific topic.

Perform knowledge assessment: the agent should evaluate how much trust the citizen should place in the knowledge discovered and make relevant recommendations. Automating the search for truth is an extremely difficult problem, as current AI approaches are discovering. Although it is possible to automate part of the trust assessment process, for example, by ranking sources in terms of reliability, complex ideas require critical thinking to distinguish fact from opinion that present intelligent machines are incapable of. Assessing the reliability of a knowledge source has to become part of a continuous dialogue between machines and humans. By iterating and learning from this dialogue, the human-plus-agent system can quickly optimize on trusting or mistrusting sources of knowledge.

Translate: in cases where knowledge acquisition needs to be translated and also when members of citizen groups or the assembly speak different languages. This is an area where standard AI approaches in language translation could deliver an adequate outcome.

Conversational intelligent agents could also be used at the group or assembly level to perform the following tasks:

Empathic monitoring: sense the feelings of citizens and be able to feed back to the deliberating process an indication of polarization versus consensus.

Performance of constitutional checks: flag proposals that may be contrary to constitutional law or in conflict with other laws.

Timekeeping: ensure that the deliberation process can progress on the basis of agreed time intervals and permissions.

Citizen assemblies have the potential to help the evolution of liberal democracies toward a more inclusive and participatory model of government. Cybernetic AI has the capability of accelerating this journey by delivering cost-effective solutions for citizen assemblies at scale. Nevertheless, there may be a much greater prize than this in our quest to reimagine democracy in the era of intelligent machines: to extend cybernetics in the way intelligent machines connect and interact with human society in order to solve the current problems of exclusion, surveillance, and marginalization.

DEMOCRACY AS A CYBERNETIC SYSTEM

There are many ways of thinking about a democratic polity, as described and discussed in various political science books over many centuries. But what if we thought of such a polity in cybernetic terms? To do so, we must agree at the outset that the difference between democracy and authoritarianism is the degree of freedom that citizens have, and that in a democracy citizens are freer. Of course, the problem with freedom is that it—by itself—does not guarantee positive outcomes. For example, people may freely elect the wrong leaders—as history has repeatedly demonstrated—and those leaders may lead the people to ruin. Therefore, what distinguishes a successful free society from an unsuccessful one is how the citizens make decisions in order to govern themselves—that is, their system of government. Governance in society is a complex matter that includes legal, procedural, institutional, and ethical aspects; however, it is also at the heart of cybernetic theory. Indeed, cybernetics' purpose is the design of successful governance in complex systems. So let us test whether it makes sense to transfer some concepts from cybernetics into the discussion about the future of liberal democracy.

A key concept worth exploring is a special state of equilibrium in complex systems called *homeostasis*. Again, the image of the flock of starlings is useful as a depiction of this special state. At any moment, the flock looks as if it is about to dissolve into its constituent parts, with single birds flying independently in all directions. And yet, the birds somehow regroup, and the flock continues its magnificent, fluid dance in the sky. This perilous dance at the edge of chaos is what homeostasis, a key characteristic of complex system behavior, looks like. Cybernetic systems exhibit such resilient behavior by using multiple feedback loops to reduce entropy and prevent themselves from decomposing under the second law of thermodynamics. Entropy is the degree of disorder in a system; the higher the entropy, the greater the disorder. The second law of thermodynamics states that the universe has a tendency to increase entropy. We know that from experience too; when a plate falls on the floor and breaks, it is impossible to put it back together as new. Indeed, as the cosmos expands and the degree of disorder in the universe increases, the ultimate destination of everything is the so-called heat death. The only resistance to the formidable second law of thermodynamics is life. Living biological systems are special because they manage to remain ordered and reconstitute themselves continuously by reducing their entropy. Our living bodies are in fact cybernetic systems in a state of knife-edge equilibrium—homeostasis—fighting death at every moment. If homeostasis breaks down irreversibly because of a trauma or a disease, the second law of thermodynamics wins and we die, that is, our bodies begin to decompose into their constituent, molecular, disorderly parts. For cybernetic systems to exist and achieve their goals, they must remain in homeostatic equilibrium.

By observing homeostatic systems, the Chilean cyberneticians Humberto Maturana and Francisco Varela suggested that self-organizing systems are also self-creating; they are, as they said, "autopoietic."[21]

Autopoietic systems are self-referencing systems that create the stuff that makes them; take, for example, a biological cell in which enzymes create the proteins and proteins create enzymes. Maturana and Varela also showed that autopoietic systems *evolve* and thus provide us with an excellent theoretical framework for understanding biological evolution, but also evolution in more general terms—for example, personal, economic, or social. Autopoietic systems can be "first-order cybernetic systems" when humans are the observers of the system—when, for example, a biologist observes a living cell through a microscope. But when humans become part of the system, they become "second-order cybernetic systems," where the observer and the observed become one. The most important difference between first- and second-order cybernetic systems is that the latter have greater freedom and autonomy—they can set their own goals. This is a very significant difference. We humans are a second-order autopoietic system: self-sustainable and capable of self-realization as well as setting our own goals, given our consciousness and free will.

Applying this theoretical framework to democracy, we can think of a democratic polity as a system of government made up of conscious humans that act together as a second-order autopoietic supersystem: where citizens are both *the subjects* that deliberate and *the objects* of deliberation.[22] This circular relationship between cause and effect, subject and object, and the parts (citizens) and the whole (assembly) is characteristic of a cybernetic system. As a second-order cybernetic system, an ideal democracy should have the ability to evolve and adapt in an ever-changing environment by achieving homeostatic equilibrium (i.e., consensus) around a common goal. An example of such a common goal would be to maximize the production and distribution of social goods by selecting among a number of policies.

Reframing the dynamics of a democratic consensus using cybernetic terms is useful not only because it gives us new insights into the nature of democratic politics but also because it allows us to take the next step and ask this question: *How can a democracy evolve*? And evolve it must because the environment—geopolitical, economic, climatological, and technological—never stays the same. New threats will always challenge the survival, cohesion, and economic development of a democracy. New technologies will be invented that will present new risks and challenges, as well as new potential for growth. Citizens will grow old and die and be replaced by younger generations who may think and feel differently than their parents and grandparents. And, of course, citizens must evolve individually too, become more responsible and self-confident by living the citizenship experience of a self-governing polity.

We saw how socialist Chile used cybernetics in the 1970s in order to achieve its economic and social goals. Cybersyn was a learning system that was embedded in the part of the socialist polity that was responsible for economic planning. The governance of Cybersyn was not dissimilar to how oligarchies use intelligent information systems today: collecting data from the edges and centralizing decisions at the core. This hub-and-spokes model of Cybersyn reflects a political philosophy that prioritizes expertise and centralized power. Cybersyn is therefore a good fit for digital authoritarianism. For liberal democracy we need a different architecture for cybernetic systems: they need to be decentralized and look more like a network of equal nodes. Such decentralized, cybernetic, networked democratic models of governance can then be embedded into the institutions of a liberal democracy in order to provide the necessary collective resilience and ability to evolve, while respecting individual freedom, dignity, and independence.

Decentralization has been an inspiration to technologists for quite a while. The philosophy behind the Internet was, arguably, pushback to the centralization inherent in monolithic computer systems and an attempt to develop alternative computer architectures that could not be controlled by a central authority. That philosophy made the Internet more resilient to catastrophes and made it a platform for enhancing democracy. But the evolution of the Internet in its web 2.0 iteration has led to disappointment. Instead of democratic dialogue, we have ended up with echo chambers of hate speech; instead of more economic diversity, we have found ourselves dependent on an oligopoly of supersized tech corporations. Thankfully, there is now new hope. For out of the debris of the global financial crisis a new technology was born with the potential to revolutionize computing and the digital economy, as well as to provide a technological platform on which cybernetic models of decentralized, democratic governance could be built.

8

THE WEB OF EVERYTHING

On January 3, 2009, the first bitcoin block was mined, marking the creation of the first cryptocurrency. It was time-stamped by its creator, the mysterious Satoshi Nakamoto,[1] who appended in the source code a curious comment: "*The Times 03/Jan/2009 Chancellor on brink of second bailout of banks.*" By including in the "genesis block" of bitcoin the headline of that day's edition of *The Times* newspaper Nakamoto was making a political statement: that this new technology—to be henceforth known as "blockchain"—was a direct challenge to the undemocratic powers of central banks. As the apotheosis of the cypherpunk movement,[2] bitcoin was raising the flag of digital anarchism, launching a techno-political revolution fueled by the outrage of the Great Recession. An alternative path to money, one that circumvented the money-printing monopoly of central banks, was now possible. Nakamoto was opening this path to the world by having solved the so-called *double-spending problem for a digital currency*.

Here's what that means. Keeping physical cash in your pocket makes paying straightforward. You walk into a store, buy goods worth $10, pay with your $10 bill, and walk out having completed and settled a cash transaction. The $10 bill you had in your pocket is now inside the store's till. But this simple transaction is not as straightforward in the digital world. If you wanted to pay for the same goods by sending $10 electronically from your computer to the store, there is an obvious problem: how would the store owner know that you were honest enough to delete your $10 bill from your computer and not keep it so you may use it again to buy something else too, that is, to "double spend"? Until bitcoin came along, the only way to ensure honesty in digital money transactions was via a third party, most often a bank. The bank intermediates between a buyer and a seller in order to guarantee that the $10 of the buyer was debited at his or her account before being credited at the seller's account. For this to happen the $10 must always "sit" in the bank's computer, while both the buyer and the store owner must trust the bank to ensure that the transaction takes place. Nakamoto's invention made such intermediation unnecessary. Owners of bitcoin could now transfer their money to anyone directly, without the need of an intermediating third party, and without anyone doubting their honesty. Once a bitcoin is spent, it cannot be double spent. It was a historical breakthrough that made many hitherto unimaginable things possible.

INTERMEDIARIES ELIMINATED BY MATH

By solving the double-spending problem of digital currency, Nakamoto invented a technology that can deliver secure exchanges where peers can buy and sell anything—money, houses, gold, loyalty points, digital cats,[3] whatever. He thus demonstrated how the Internet of Information could be transformed into the Internet of Assets: peer-to-peer exchange need not be limited to information anymore but can extend to include anything in the real, as well as the imaginary, world. Moreover, blockchain allows for the near limitless *fractionalization of assets*; for example, you can sell or buy tiny percentages of the value of a painting, a house, a spaceship, or a patent. Fractionalization of assets brings down many barriers to private ownership and opens up new possibilities for capital creation to all citizens, rich or poor.

So what is a blockchain? The simplest way to imagine it is as a ledger that records transactions between participants in an interconnected computer network. This ledger is not a central database, as is the case in most classic software applications, but is instead *decentralized*, meaning that every participant in the network has a copy of the most updated version of the ledger on their computers at any time. This way, all transactions taking place on a blockchain—which are, effectively, electronic messages sent to all participants—are fully transparent to everyone. The practical way that this takes place is as follows: transactions are bundled into "blocks" over a set period of time and are then *verified* so they become official records in the ledger, like blocks strung together on a chain. Once that happens, all the nodes in the decentralized ledger are updated with the new block. The critical step in the process of recording new blocks is verification. Without "objective" verification, none of the participants could trust that the transactions recorded in any given block were not fraudulent. But how can this verification take place without a "trusted" third party? How can participants in a blockchain be kept honest without some higher authority policing their actions? There are two important mechanisms that ensure that decentralized, or leaderless, verification takes place.

The first mechanism is the cryptographic *method of verification*. There are various methods, some already used and others in experimental phase. The problem that all those methods aim to solve is practically the same: how can one know that any given node in the blockchain network is telling the truth when verifying a transaction? This problem is known in computer communication theory as the "Two Generals Problem," and is easily explained using a simple story. Two generals, one commanding and one following, attack a common enemy from opposite directions. However, each general's army on its own is not enough to defeat the enemy. The only way to win is for the two generals to cooperate and attack at the same time. To cooperate, they need to send messengers across the enemy camp. Messengers may be captured and the message not delivered, whereby the attacking general will be defeated. Can the generals trust that their message was received? Using game theory, the problem was shown to be unsolvable; alas, there is no way for any general to be certain that their last messenger had safely passed through the enemy camp.

Nevertheless, in 1982 a generalized version of the Two Generals Problem was published,[4] and the problem was renamed as the "Byzantine Generals' Problem."[5] In this new story there are more than two generals, but some are traitors. Messages are sent between generals to coordinate an attack. Traitors would do their utmost to corrupt the messages and spread lies in order to foil the attack. What needs to be done in order to ensure that the correct messages are sent and received to every general, regardless of the traitors' efforts? This general problem was shown to have a solution: the generals can reach a *consensus* about the truth of a given message by taking a majority vote. The mathematical algorithm that solves the Byzantine Generals Problem is shown to reach consensus—that is, to give either a "true" or "not true" output—as long as two-thirds of the generals are honest. If the traitors are more than one-third of the total, consensus is not reached, the armies do not coordinate their attack, and the enemy wins.

The Byzantine Generals Problem applies to a blockchain when the truth of a transaction needs to be verified. Any peer on the network can act as a treasonous Byzantine general and transmit a false transaction to nullify the blockchain's reliability. Given that there is no central authority to take over and repair the damage of false messages, the network itself must be "fault tolerant"; that is, the network must run some code (or "protocol") that solves the Byzantine Generals Problem every time a message is transmitted, so that consensus is reached and truth is verified. In the case of bitcoin the solution to the Byzantine Generals Problem is probabilistic and is called "proof of work." This method creates a high cost to anyone who verifies a transaction by demanding a solution to a mathematical problem. This cost accrues because of the computing that is necessary and the amount of electrical energy that must be spent in order to solve the mathematical problem.

The second mechanism of verification is economic *incentivizing*: whoever bears the cost of verification should be motivated to do so. In the case of bitcoin the verification of a transaction takes place by so-called miners. Every time bitcoin miners verify a block, they get rewarded for their effort with a number of newly minted bitcoins,[6] just like the real miners of yesteryear who put forth effort and sweat digging for gold.[7] This way they compensate for the cost they incurred in order to verify a block, and they make a profit too. Incentivizing with bitcoin rewards, as well as additional transaction fees paid to participants for offering other services, ensures that a blockchain is self-sustaining.

Because of their decentralized nature, blockchains are generally more secure than centralized systems because they do not have a central point of attack or failure. The consensus mechanism is another key security feature that ensures that everyone follows the same rules, and that everyone agrees on the state of the network at any time. In addition, data integrity is guaranteed because blockchains are "immutable": the protocol that adds transaction blocks to the chain must verify that the new block's "truth" is dependent on the verified truth of all the blocks that preceded it. This mechanism prevents alteration of transactions already confirmed. If a malicious hacker wanted to falsify a transaction, he or she would need to spend impossible amounts of effort to retro-falsify every transaction on every block in the chain. In such a scenario, conflicted or false copies of the ledger would be quickly eliminated through the sheer weight of the math involved in mining.

Another important property of blockchain is the built-in protection of participant *anonymity*. Each transaction on a blockchain is associated with a private key that belongs only to individual actors participating in the network. This private key is an abstract series of numbers and characters and does not reveal the identity of the owner. To make a transaction on a blockchain network, a participant uses this private key. It is therefore impossible for a third party to identify who is transacting. This characteristic has triggered much negative reaction from governments and tax authorities across the world. Anonymity on a blockchain seems to present an ethical dilemma. On the one hand, it protects criminals from facing justice and the law. But, on the other hand, it also guarantees privacy and prevents citizen surveillance by an authoritarian state. On closer examination, however, this ethical dilemma can be shown to be less so because of the potential to embed lawful contracts in the software protocol that verifies transactions on a blockchain. These contracts can include a condition to verify one's legal identity before being allowed to make transactions. This condition may not compromise anonymity but simply return a yes-or-no answer for verification purposes. To understand how such "smart contracts" could enable new ways of decentralized, leaderless human collaboration, we must take a dive into an emerging field of economics for blockchains called "cryptoeconomics."[8]

THE TOKENIZATION OF EVERYTHING

The basic principles of cryptoeconomics are relatively simple. Blockchain—also referred to as distributed ledger technology—is used to build an encrypted network for willing participants called a cryptonetwork that provides some service. It could be any service—for instance, data storage, or access to computation facilities, or a marketplace for work or for trading securities and digital or physical assets. Participants in this network are incentivized via the issue of a "token." These tokens are written in software code as smart contracts, which means that they can be anything their designer decides; for instance, they can be a currency, a commodity, and a security all-in-one. They become valuable because of the consensus mechanism used in verification and also because participants recognize that using tokens on the cryptonetwork is fully secure. Participants transacting on the cryptonetwork may use this token for their exchanges. Depending on how they contribute to the function of the cryptonetwork, they are rewarded with tokens. For example, some participants may contribute by verifying blocks (similar to the miners of bitcoin), others may write or improve the code, provide marketing and community management services, or provide general support. The governance of a cryptonetwork can be leaderless and automated by the source code of the blockchain. The designers of cryptonetworks can formulate specific governance processes; for instance, participants who wish to be part of a decision-making body may "stake" their tokens to participate in committees. This way each cryptonetwork can also have its own form of governance.

Cryptonetworks can be both private and public.[9] Private means networks controlled by a single authority, where the source code running the network is proprietary and nodes in the network need to ask permission in order to join. Corporations use private cryptonetworks to automate business processes where the costs of agreeing on a "single version of truth" are high. Typical use cases of private cryptonetworks include syndicated loans or logistics. In such cases many businesses need to cooperate in a process where much time and effort is spent in comparing documents and verifying transactions. Blockchain can provide the means to accelerate these processes and significantly reduce costs. Think, for example, how shipping companies move goods from one geographical area to another, often needing to hand over from one transport agent to another, ensuring compliance with tax and customs laws in a variety of countries, purchasing insurance, making checks on insurance and safety policies, and so forth. All these checks can be automated using a blockchain where all the participants in the process have access.

But the really disruptive power of blockchain is probably in public, or "open," cryptonetworks; where the entire source code is available to everyone under an open license and there is no central authority to control access. These truly decentralized cryptonetworks are open to anyone, and anyone can join as a node and provide some network service without asking for permission. Cryptoeconomics uses game theory to design and implement economic incentives for leaderless human collaboration on open, permissionless cryptonetworks. Cryptoeconomics can thus enable the realization of new types of peer-to-peer digital platforms that use blockchain technology—called "cryptoplatforms"—where transactions between participants are effected through the exchange of tokens. Cryptoplatforms and tokens have the potential for solving one of the biggest problems in the accelerating digitalization of our economies, namely, the *wealth asymmetry problem*, because they are disrupting the dominant economic model of the digital economy.

WEB 3.0

As discussed earlier, the evolution of digital platforms over the past ten years has given rise to the gig economy by aggregating, dispersing, and automating work. The gig economy is leading liberal democracies toward greater income and wealth inequality, especially as more and more work tasks are performed by automated systems that learn from our data. Digital platforms were made possible by the convergence of three technologies over the past ten years: cloud computing, social media, and mobile technology. These technologies gave birth to "web 2.0," or the "social web." As we look into the next ten years there are three new technologies that will shape the digital economy: AI, blockchain, and the Internet of Things. These technologies are enabling a new version of the web called "web 3.0"—or the "web of assets"—a place where not only information but also *everything* can be exchanged.

Web 3.0 exchanges of everything could continue to benefit the few, but there is also the potential to enable ways to distribute the enormous bounty of the AI economy more equitably and diffusely. To understand how this may be possible, we need to examine the differences between web 2.0 closed platforms and web 3.0 open cryptoplatforms.

What we have witnessed in the short history of the digital economy so far is that the owners of web 2.0 digital platforms—the founders, the executive teams, and their investors—have become richer while the users on those platforms have not. This seems to be unfair. For without users—the buyers, sellers, evangelists, reviewers, riders, drivers, and so on—those digital platforms would be worthless. Web 2.0 platforms are valuable only as long as we, the users, use them. If no one got a ride via Uber, or no one wanted to be an Uber driver, the stock of Uber would collapse overnight, despite all the great talent, software, and investment that has gone into building this amazing company. If no one wanted to post content on Facebook, or use Google to search the web, those companies would also disappear. It is the "liquidity" of a digital platform, measured by the volume and frequency of user interactions and transactions, that makes a platform valuable. However, most of this generated value goes straight to the founders and investors, who, thanks to their expert lawyers and tax advisors, make sure governments get little by way of tax—as the European Union has recognized and thus started pushing for a new regime of "digital taxation"[10] fit for the digital era.

Figure 8.1 illustrates the misalignment of goals between owners and users (or "participants") in the existing digital platform economy. Because of this misalignment, platform owners care for their users only when the participation acquisition costs—the so-called customer acquisition costs (CACs)[11]—are high. That happens during the initial period following launch, when a platform needs to invest massively in creating both demand and supply at the same time. But once enough capacity has been built on both sides of the platform, the participation acquisition costs begin to drop. Users are attracted to the platform because there are already transactions taking place. As more participants are attracted, the transactions become more liquid, a phenomenon called "network effect." As the platform grows, network effects lower the customer acquisition costs even further, and platform owners start caring less about individual

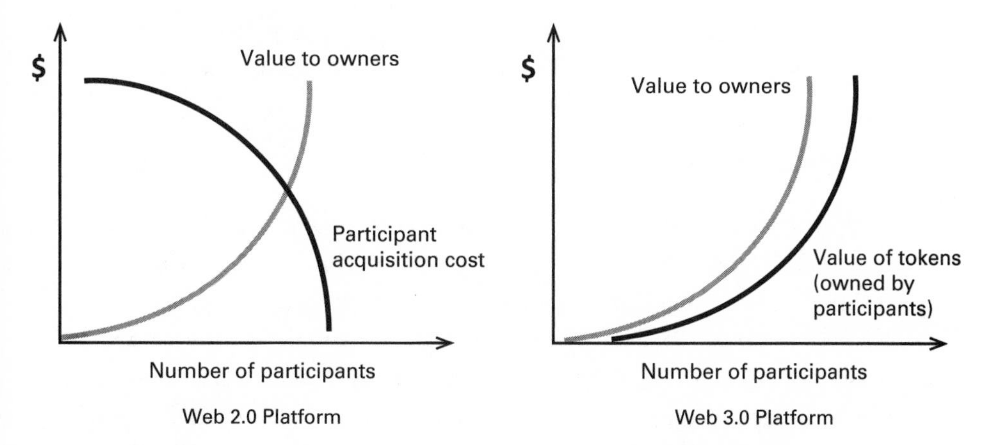

Figure 8.1 Comparing web 2.0 platforms to web 3.0 platforms. ©2019 by George Zarkadakis.

participants. Web 2.0 platforms are typical rent-extraction machines: the value created by the participants is funneled to the owners. But thanks to the unique properties of cryptonetworks—or "web 3.0 platforms"—all this could change.

Cryptonetworks enable shared ownership of generated value via the wide distribution of tokens among a platform's participants. As can be seen in figure 8.1, as the liquidity in a web 3.0 platform increases, the value returned to the owners increases too, but so does the value of the tokens held by participants. Tokens can take various forms in order to incentivize participation. They can be tokens that act as currency to be used in transactions, but they can also be tokens of equity options, or shares on dividends, or indeed anything else that innovative entrepreneurs in the era of web 3.0 may come up with.

Services running on open public cryptonetworks shift the power balance in favor of users, and from the few to the many. Imagine, for example, if a number of Facebook's users disagreed with the way Facebook is evolving its services or behaving in public. In today's world, they could protest only by writing a few blog posts. But a Facebook written as an open-source cryptonetwork allows for "forking," which means participants can use the same code to create a similar service governed by their own, preferred rules. In a world of public, open-source blockchains, users could vote with their feet and create Facebook 2.0 without facing legal repercussions. As we will examine later, public cryptonetworks can be applied to enable new forms of company organization and governance.

CRITICISMS OF BLOCKCHAIN

Cryptocurrencies have been notoriously favored by criminals and tax fraudsters, while many "initial coin offerings" (ICOs)[12] were shown to be nothing more than get-rich-quick scams. Bitcoin started trading in October 2012 at US$0.125 and, by December 2018, had risen to the vertigo-inducing heights of US$19,783.21 per coin, mostly because of speculation, only to collapse soon afterward, as many economists had predicted. Since its inception blockchain has had a wild ride on the "hype" slope of the Gartner cycle,[13] making the news almost daily, with passionate camps firmly pitching their tents on both sides of the divide on what the future of crypto might be. For critics, blockchains were a fool's gold, an inefficient and costly technology that delivers little more than data structures for authorized users to add more data to them. For supporters, they are a historical breakthrough that replaces trust with lines of software, echoing Satoshi Nakamoto's statement in his original white paper: "We have proposed a system for electronic transactions without relying on trust."[14] As the stardust settles in the crypto world and the technology enters its "trough of disillusionment" stage, it is time to coolly examine the arguments of both critics and supporters.

There are two main categories of criticism of blockchain, technical and ontological. Technical criticisms revolve around the inefficiencies of distributed ledgers—for instance, the high cost and environmental impact of running proof-of-work verification protocols during mining.[15] Given their immutability, there is also the inherent problem of "append-only data structures" in a blockchain. This means that you cannot delete or edit data in a blockchain. You are only allowed to add more data to it. This is quite problematic because blockchains are a mismatch to existing business systems—for example, in banking or government, where it is vital to be able to go back and manipulate past data records. And, finally, there is the issue of security. Although blockchains claim to solve security problems by not having a single point of failure, as well as through the cryptographic hashing of transaction records, the truth is that security risks do not disappear but are only moved further upstream. For example, cryprocurrency holders are vulnerable to their cryptowallets being hacked. Something as simple as forgetting one's log-in details can also lead to losses. Exchanges can get hacked, and smart contracts may have bugs that are extremely hard to debug given that blockchains are append only. All these problems are important and crucial for the future of blockchain technology and cryptocurrencies. Unless they are solved, the opportunity of blockchain will never materialize and its application will be restricted.

Nevertheless, these are all *engineering* problems and therefore, ultimately, solvable. Blockchain technology is still evolving. Security issues are being addressed, and new and more efficient verification protocols are being tried. For instance, an alternative to the environmentally damaging proof-of-work method is called "proof of stake." Instead of miners spending heavy computing power to solve a mathematical problem to reach consensus, all participating nodes place a bet on blocks. The nodes whose block is the honest block (i.e., contains no fraudulent transactions) get rewarded. The nodes whose block turns out to be dishonest get penalized; the amount of their bet gets debited from their balance. Placing bets doesn't require high-performing computers and electricity. All a node needs to be eligible to get rewarded is some stake that it can place a bet with. The idea behind this method is, "whoever has the maximum stake in the blockchain must have the loudest voice." It is therefore not unreasonable to assume that, in time, innovation will deliver much more secure and efficient systems. Current technical criticisms will very likely disappear as blockchains enter the "plateau of productivity" in the next three to five years.

Ontological criticisms, however, that blockchains are not really "trustless," need to be taken much more seriously. Blockchain enthusiasts claim that mathematics and software code can replace trust. By wearing T-shirts with messages such as "in code we trust," they claim that the combination of a distributed ledger, a consensus algorithm, and the issuance of tokens is enough for humans to transact without the need to trust anyone else but themselves. But can software code *really* deliver such wide-ranging social trust? It seems very unlikely. There is a whole ecosystem of various technology providers and stakeholders around a blockchain system, and every participant in that ecosystem wields some degree of power or influence. Importantly, all those participants are human. Math and software can only take you so far. It seems that claims of blockchain having solved the problem of human trust are overblown and misplaced. In order to trust any system, you must trust the people who made it, maintain it, and run it.

For example, you need to trust that those who coded the smart contracts had the right skills and intentions, as did everyone else who coded every other bit, piece, pipe, and engine that makes a cryptonetwork possible. If something goes wrong—as it surely will—there must be some kind of governance structure and mechanism to decide how to fix it. The human part of governance of a blockchain cannot be fully automated using smart contracts. If it were, it would instantly become subverted by the same problems that it was meant to resolve. In other words, you cannot have faulty code fixing faulty code. Some human needs to intervene at some stage to get things sorted out. That human needs to have the authority to do so. How this authority is given, by whom, and under what circumstances are matters of governance. They are also matters of legal compliance. Blockchains cannot operate outside the law and the courts. Doing so would expose participants in a public blockchain to predation as well as to prosecution. Trust is a social good too vital and too complex to be replaced by math. Humans will inevitably remain at the core of whatever is built using this technology, as creators, participants, contributors, adjudicators, regulators, and governors. Blockchains are therefore revolutionary not because they replace trust with math but because they enable alternative forms of decentralized governance of digital platforms. This, as we will see, is key to democratizing the future AI economy.

BLOCKCHAIN FOR DEMOCRACY

Blockchain may not be so much in the news nowadays, but it is not going away either. It has already set off a wave of change that is unstoppable. Bitcoin may have lost 50% of its value at its highest point but still trades steadily around the US$7,000 mark. A market for cryptocurrencies is continuously evolving. There are countries considering diverting national energy resources into industrial-scale bitcoin mining.[16] New exchanges are being launched around the world, and forward-looking regulators are shaping new legal frameworks that weed out speculators and allow for the innovation of new, and regulated, financial products. Facebook has proposed Libra, a cryptocurrency backed by a fund of fiat currencies, setting off much furor and anxiety among central bankers and governments as to how much such a privately run currency could destabilize global money markets. It is not only private corporations that see the opportunity for blockchain to revolutionize the financial world. In late 2019, Mr. Mu Changchun, the head of the digital currency research institute of China's People's Bank, told a forum in Hong Kong of the bank's plans to adopt a digital currency, powered partially by blockchain technology, called "Digital Currency Electronic Payment."[17] A few months earlier, in August 2019, the governor of the Bank of England, Mark Carney, in addressing the annual meeting of the Federal Reserve Symposium in Jackson Hole, Wyoming, reiterated his support for a global, central bank–issued cryptocurrency called "synthetic hegemonic currency" (SHC), perhaps through a network of digital currencies. This was a very bold proposal, for, if SHC were to happen, it would cause a major paradigm shift in the global financial system, effectively deposing the US dollar as the world's reserve currency. Carney was quite explicit on that, adding in his speech, "An SHC could dampen the domineering influence of the US dollar on global trade. If the share of trade invoiced in SHC were to rise, shocks in the US would have less potent spillovers through exchange rates, and trade would become less synchronized across countries."[18] Carney was right to be concerned about de-risking the instability of financial markets in an age of renewed trade wars, since such instabilities directly impact the politics in liberal democracies and the welfare of citizens. But blockchain technology could facilitate the reinvention of our political system and our economies in many ways beyond the adoption of a global—and more stable—cryptocurrency.

Throughout this book I have argued that for liberal democracy to survive in the age of intelligent machines we must solve the problem of current, and future, wealth and income inequality. As AI systems automate work and we and our children become part-time workers at best, or chronically unemployed at worst, we must find ways to counterbalance the loss of a steady income and the personal and family security that this income brings. I have also argued that universal basic income is not only fiscally problematic but also insufficient to providing a decent living standard—let alone supporting the accumulation of capital and wealth by the many. It is also in conflict with classical liberal values of a minimal state, as it requires an even greater role for government as a redistributor of wealth. We therefore need noncoercive, bottom-up ways to more evenly spread the new wealth that AI will create *and* deliver true prosperity rather than mere sustenance.

For the wealth that AI will create will be tremendous: a 2018 report by management consultancy firm McKinsey estimated that AI could deliver an additional US$13 trillion to the global economy by 2030.[19] That is almost a 15% boost in today's economic output. This prediction is in line with similar forecasts made by Accenture and PwC. Almost everyone agrees that the dividends from adopting AI in business and government will be enormous. Nevertheless, in the same report, McKinsey also warns that the "adoption of AI could widen gaps among countries, companies and workers."[20] Continuing with "business as usual" is clearly not an option, for it will exacerbate the causes of popular mistrust in liberal democracy as a system of government that enables a fairer distribution of wealth. We need radical change so that everyone has a stake in the future AI bounty. As I have argued, one of the reasons for so much inequality in a digital economy is that citizens are not earning anything in return for their contribution to the success of digital platforms, as all this value is extracted from the platform for the benefit of a few. Cryptonetworks, by introducing a new, decentralized way of governing digital platforms, could be the way to democratize their governance and give citizens who use them, or otherwise contribute to them, a say in how they are run. Given their tokenization capability, cryptonetworks also hold the promise that the new web 3.0 could be a place for spreading the wealth of the Fourth Industrial Revolution more diffusely. This is now a more realistic prospect than ever before because the AI economy is largely based on our data. With blockchain technology, as we shall see in the next chapter, we can trace and record the value that each individual piece of data contributes to the success of a digital platform that is powered by AI algorithms. In other words, we can account for the contribution that each one of us makes, and we can use this as a monetization and repayment mechanism.

Furthermore, a technology that delivers a decentralized way of governing human collaboration can be leveraged to experiment and develop new forms of business organizations that have been hitherto impossible. Just imagine a leaderless, virtual company in which millions of people contribute through their data, ideas, and work in collaboration with intelligent machines—a cybernetic learning system, such as the ones discussed in the previous chapter, wherein the goal would be the production of a high-value social or economic good. Our imagination is the only barrier to how we may leverage technologies we currently have at our disposal in order to reboot capitalism in the twenty-first century and deliver a new iteration of a free, democratic, and prosperous society.

9

DEMOCRATIZING THE AI ECONOMY

Since the global financial crisis, numerous debates and discussions have taken place on how to reform the market capitalist system of free trade, globalized financial markets, and privately owned corporations so that it respects the preservation of the natural environment and the welfare of local communities, promotes social equality, and produces not only profits for the few but also social goods for the many. More recently, the Business Roundtable, a business group of more than 200 members, including the chief executives of JPMorgan Chase, Amazon, and General Motors, and of a combined annual revenue of US$7 trillion, published a statement of purpose calling for a significant departure from the dogma that businesses should serve only the owners of capital.[1] Companies should "protect the environment" and treat workers with "dignity and respect," the statement said. Without doubt there is growing awareness in corporate America and elsewhere that Milton Friedman's philosophy of businesses caring only for maximizing value for their shareholders is outdated. We now live in a different world. The threat to life on our planet due to climate change requires corporate leaders to think long term and not just of the next quarterly results. Millennials and younger generations are more sensitive to issues of corporate responsibility and sustainability and often boycott brands seen to be all about healthy balance sheets and little else.[2] Moreover, there is considerable pressure for reforming corporate purpose from politicians, especially as corporate profits soar and many companies opt to spend those profits on share buybacks instead of improving worker wages.

And yet, there are limits to what governments in liberal democracies could do. They could certainly regulate executive pay more and rebalance tax systems so that corporate profits are taxed more while work is taxed less. They could introduce smart policies for dealing with technological disruption, for instance, by using tax incentives to promote technologies that generate new and more high-value work for humans rather than simply automating tasks, and by future-proofing health care, pension, and welfare systems by taking into account the impact of automation technologies on workers. Such measures and policies would surely smooth the impact of intelligent machines in our societies over the next few years and would quite possibly reestablish some degree of citizen trust in democratic institutions.

Nevertheless, as I have argued throughout this book, we also need to think beyond what governments can traditionally do to regulate markets, and explore more radical approaches to reforming capitalism in the Fourth Industrial Revolution. We need to think out of the box in order to deal with the incoming disruption, both because of the enormity and novelty of the challenges that await us and also because of the unique opportunities provided by intelligent technologies. In this chapter I would like to focus on two such opportunities. The first is how we can go several levels deeper from the macroeconomics of capitalism and reimagine some of the fundamental elements in the machinery of corporations. More specifically, I will explore how the governance of digital platforms can be redesigned in order to give birth to new types of business organizations that can have purposes and goals beyond profitability. Second, I will look into the "new oil" of the AI economy—namely, data—and discuss how we can create new, game-changing institutions and data-driven paradigms that could transform every person in the world into a value-creating asset.

REINVENTING BUSINESS GOVERNANCE

Companies are not only economic entities but political entities too; their organizational model and method of governance reflect the zeitgeist of their times. As the first industrial revolution took hold in Victorian Britain, joint-stock companies acquired the principles and language of representative democracy. Shareholders were often called "constituents."[3] Like a parliament, general assemblies of shareholders would elect directors—the "executive government"—and collectively decide on important issues using a voting system based on the number of shares. According to an observer in 1807, the General Court of the East India Company was akin to a "popular senate," with the important difference that voting rights did not discriminate between the sexes.[4] Robert Lowe, a nineteenth-century British statesman, called these companies "little republics."[5] The direct participation of shareholders in a company's governance began to diminish as company operations expanded around the globe and shares started trading over the counter of stock equity exchanges. Thus, the little republics of nineteenth century early industrial capitalism[6] regressed to dictatorships run by omnipotent boards of directors who ruled over strictly hierarchical organizations. As stock markets grew in size and importance, shareholders could now buy and sell shares of any publicly listed company, anytime, and so they lost interest in the purpose of the companies they invested in. Shareholder interests also diverged from the interests of employees, suppliers, customers, and local communities where those companies were based. The long termism and plurality of goals that characterized early joint-stock companies was gradually replaced by the short termism and the singular purpose of quarterly financial performance.[7]

The advent of the Internet in the late twentieth century set off the transformation of the global economy into a "digital economy" and, as discussed earlier, enabled a new business model to emerge, the digital platform. Google, Microsoft, Apple, and Amazon are companies harvesting the value that is created and added by participants interacting and transacting on their digital platforms. They are also the most valuable companies in the world today based on market capitalization.[8] But the digital economy did not disrupt corporate governance. Digital companies have retained the characteristics of "enlightened dictatorships" run by visionary CEOs and boards of directors. Centralization of decision-making is still at the heart of corporate governance in digital, as well as less digital, companies today. Without doubt, centralization confers important operational efficiencies, but it can also promote groupthink

	Industrial Era Hierarchies	Web 2.0 Platforms	Web 3.0 Cryptoplatforms
Mindset	Command and control	Orchestration	Collaboration, empower individuals and communities
Core Assets	Ownership of financial and physical capital, employees, IP, brand	User networks, proprietary platform software	Data
Value creation	By the company	P2P and via platform capabilities	P2P
Value capture	By the company (shareholders)	By the platform (shareholders)	By participants
Strategy and knowhow	Well-established (100+ years)	Moderately established (20+ years)	Nascent
Use of blockchain	Technically possible in certain cases with unclear advantage	Yes. Focus on tactical advantages, cost, and efficiency	Yes. Strategic.
Governance	By company	Platform rules and owning company	P2P, community-based
Reputation	Brand name	Curation, rating systems, data	P2P
Security	Vulnerable, central point of control	Vulnerable, central point of control	Blockchain consensus
Privacy	Regulated by national laws	Regulated by national laws	Blockchain protected
Regulation	Well established	Nascent (GDPR)	Nonexistent
User data	Owned by company	Owned by platform	Owned by participants

Figure 9.1 Comparing web 2.0 platforms with web 3.0 platforms. IP, intellectual property; P2P, peer to peer; GDPR, General Data Protection Regulation. ©2019 by Vince Kuraitis, reproduced by permission.

and hinder agility, innovation, and resilience to change. What concerns us here is how the current, centralized governance model in most corporations exacerbates income and wealth asymmetries, particularly in light of advancing automation due to AI—and whether there is an alternative.

As discussed in the previous chapter, web 2.0 platforms are profoundly different from web 3.0 cryptoplatforms. The former are essentially centralized rent-extraction machines, while the latter can be decentralized and widely distributive of the economic value they generate. Figure 9.1 summarizes the main differences between the two business paradigms and compares them to Industrial Era hierarchies.

Web 3.0 cryptonetworks[9] can be peer-to-peer networks that resemble "mutual organizations," such as cooperatives or mutual insurance companies, in that members derive rights to profits through their participation as workers as well as customers.[10] Cryptoeconomics provides the theoretical framework and blockchain the technology to deliver a more democratic governance of digital platforms on web 3.0. Participants in such platforms may earn tokens for their participation, use those tokens to exchange economic value, and also store them as an asset of value that grows as more exchanges take place across the network. Thus, the network effect that results from participants' contribution returns some of the value created directly back to the participants. Because tokens are lines of code, there is virtually no limit to the degree of innovation that can occur with respect to ways of incentivizing participants on web 3.0 cryptoplatforms.

Tokens can also be used for voting on matters of governance. For instance, in proof-of-stake protocols, participant voting power is proportionate to the amount of tokens each participant has. This approach has power balance consequences when a small majority may end up controlling more than 51% of stakes and imposing its will on the rest. In such cases the minority may disagree with the way that the platform is being developed or managed. In a centralized governance world, a minority of shareholders would have a very limited set of options to counteract an authoritarian majority; they can protest, go to the courts if they can afford it, or sell their stock and bear the losses. But in a decentralized cryptonetwork, a disenfranchised minority can do much more: they can make a hard fork and create a new blockchain. Forking, by being the "nuclear option" of dissenting minorities on a web 3.0 platform, provides the ultimate inhibitor to a controlling majority becoming authoritarian, for if they did, they would run the risk of many valuable participants in the network simply walking away and becoming competitors, which would then lower the financial value of the network.[11] Forking is therefore a mechanism for realigning the interests of all participants, regardless of the varying degrees of voting power that they may have.

The crucial interplay between the balance of political power on a cryptoplatform and the economic value that the platform can produce requires a layered approach to decentralized governance (see figure 9.2). At the most basic layer, the founding stakeholders must agree on a set of fundamental principles, a kind of "constitution" for the platform. It is at this constitutional layer that human interaction and deliberation can make use of the citizen assemblies' model. Founders may come together

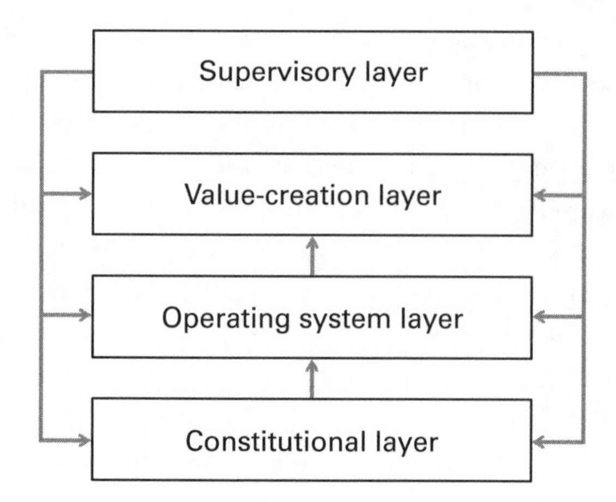

Figure 9.2 A layered approach to embedding democratic governance in web 3.0 cryptoplatforms. ©2019 by George Zarkadakis.

from a range of stakeholder groups, such as workers, software developers, hedge funds, foundations, local governments, entrepreneurs, and so on. Those founders will have to first agree on the purpose of the platform, its business objectives, and its values and ideals; for example, they may decide to safeguard jobs, offer health insurance, pensions, or childcare, or protect natural resources. The constitutional layer must also be compliant with national and local laws and regulations. By democratizing the process of defining the foundational constitution, a web 3.0 cryptoplatform can be established as one of those "little republics" that Robert Lowe was describing in the nineteenth century—or indeed as a Cyber Republic of the twenty-first century!

Once the foundational constitution has been decided on, software developers can start building the second, derivative layer of the cryptoplatform, where smart contracts will track, audit, and regulate exchanges on the basis of what the constitution allows and promotes. This would be the operating system layer of the cryptoplatform, where the mechanisms of recording of transactions and tokenization will be built. On top of this layer, other developers will then build the distributed applications that will run on the platform and produce economic value for everyone, thus creating the value-creation layer. Finally, an additional supervisory layer is needed to resolve issues that are certain to occur as the cryptoplatform evolves over time. For example, at this layer, stakeholders will be able to

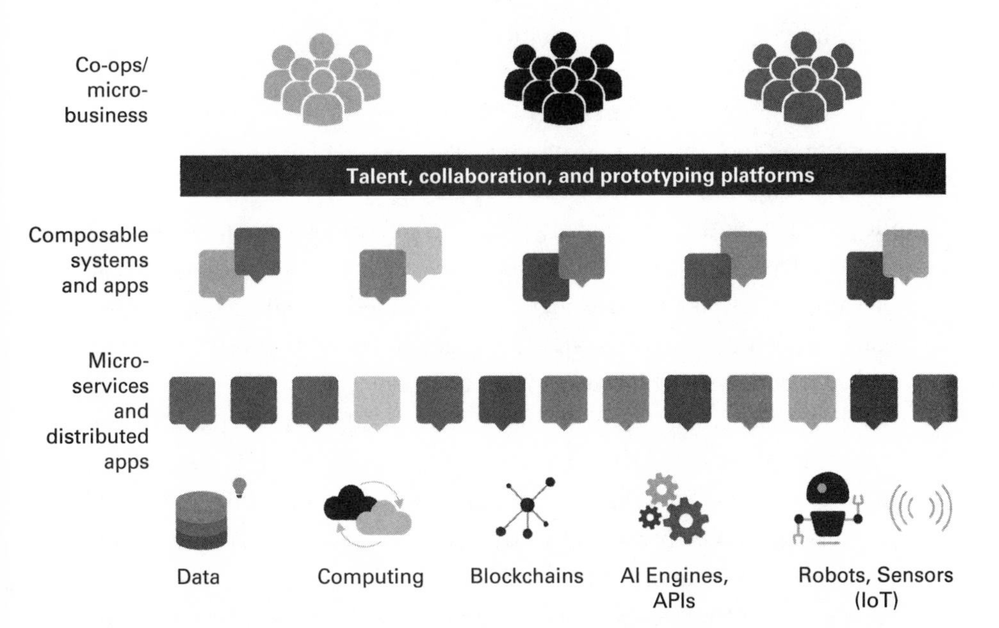

Figure 9.3 Web 3.0 ecosystem. Interconnected cooperatives and microbusinesses collaborate by leveraging a stack of distributed and decentralized technologies. AI, artificial intelligence; API, application program interface; IOT, Internet of Things. ©2019 by George Zarkadakis.

address and vote on resolving technical issues and conflicts, as well as making complex decisions. Although the foundational principles of the cryptoplatform will have been decided in the cryptoplatform's constitution, the supervisory layer must also allow, in extreme cases, the community of stakeholders to alter aspects of the constitution and propagate these changes to the other layers as well.

ORGANIZATIONS OF THE FUTURE

This participatory, layered, collaborative model of governance and codevelopment can be used to imagine new types of business organizations. Figure 9.3 shows how business ecosystems on web 3.0 that are decentralized and democratic in their governance could function as a "stack" of services that may themselves be delivered by networked cryptoplatforms. These ecosystems would connect a variety of businesses including cooperatives, traditional corporations, or other "microbusinesses" (e.g., small teams of professionals).

In such a future scenario, web 3.0 ecosystems may connect to other ecosystems, too, and thus create a fractal economy of interconnectedness and cooperation facilitated by networks of open blockchains. For example, take a hypothetical open cryptoplatform that provides a car-sharing service. This "full stack" cryptoplatform is a self-contained ecosystem of software developers, designers, drivers, riders, social media experts, communicators, marketers, and so on, governed by its constitution and its smart contracts. This autonomous ecosystem, however, is dependent on the existence of infrastructures such as roads and the provision of fuel,[12] as well as the provision of other services such as car insurance, car maintenance, car mobility qualification, and so forth. All those ancillary services can be provided by other cryptoplatforms. Moreover, many public services, such as public utilities and road maintenance, can also be reimagined in the context of cryptogovernance, as we shall see in more detail in the next chapter.[13]

Cryptoplatforms can be autonomous *and* interconnected at the same time. And anyone can be a participant, and contribute in various roles, in more than one cryptoplatform. We can thus imagine organizations of the future that use blockchain to track the contribution of millions of participants, a type of "swarming" organization that delivers some social good. The example of Wikipedia is perhaps one way of visualizing those multimillion participant businesses of the future, the significant difference being that in those future businesses participants will also be remunerated for their contributions via tokens.

Cryptoplatforms can also be completely leaderless, as the Decentralized Autonomous Organization (DAO) on the Ethereum blockchain demonstrated in 2016.[14] The DAO aimed to showcase the feasibility of a revolutionary form of business organization that had no owners and run without the need of executive management or a board of directors. It was crowdfunded via the issue of a token and managed to raise over US$150 million from around 11,000 investors. The business purpose of the DAO was to invest in various ventures and projects and distribute the profits from those investments back to the token holders. Flaws in the computer code resulted in the DAO getting hacked, and millions of investor dollars getting siphoned away into the hackers' accounts. Ethereum ultimately managed to recover the funds, but only after making the very difficult decision to "hard fork" the blockchain. As a result of that forking, the Ethereum blockchain comes today in two "flavors," Ethereum and Ethereum Classic. Nevertheless, the DAO remains a model for future organizations that needs to be improved, but also regulated. For instance, the US Securities and Exchange Commission (SEC) has decreed that digital tokens, such as the ones traded in DAO, should be considered securities and therefore fall under the respective laws that require the registration of their owners, including compliance checks on money laundering and terrorist activities.[15] Such regulation is essential if those new types of organizations are ever to leave the experimentation stage and enter the mainstream.

A SHARING ECONOMY FOR WORKERS

Web 3.0 cryptoplatforms can also allow us to imagine gig-working platforms that serve the interests of workers and not only the interests of platform owners or investors. This is of vital importance for the future, as automation is likely to result not in the complete elimination of work but in the gradual replacement of steady, salaried jobs with part-time and intermittent employment.

To return to the ride-sharing example, take Uber, currently valued in the tens of billions of dollars. There is no reason why a group of drivers shouldn't be able to build a similar platform for themselves: the software tools to develop it are open-source and readily available, and development and design talent can be given incentives to participate. A blockchain-based ride-sharing platform called La'Zooz,[16] founded by Shay Zluf in Tel Aviv, was one such proof of concept. The blockchain protocol allowed anyone to join and begin to earn Zooz tokens each time they drove over 20 km (about 12.4 miles), or contributed code to the design of the app, or got others to join. Zooz tokens would then be used as the internal currency of the ride marketplace. Similar initiatives have sprung up elsewhere too. In Singapore, a Korean team called MVL introduced TADA, the equivalent of Uber on blockchain. The designers of TADA[17] view their application as a means to create a new ecosystem that would include drivers, riders, and automotive industry companies and adjacent service industries and their customers. In China, a company called Didi[18] is building another blockchain-based ride-sharing application called VV Go, which they plan to test in Singapore and then extend to other cities, too, such as Toronto, Hong Kong, and San Francisco.

Another example of using blockchain to reinvent the gig economy is Supp,[19] an application that connects contingent hospitality workers and those who wish to hire them. The market that Supp addresses from their headquarters in Melbourne is rather sizable; it is estimated that around 800 million families globally earn their living in the hospitality industry,[20] while according to the UN around 200 million millennials are traveling around the world, working on and off in hospitality and leisure jobs. This market is currently fragmented and subject to the web 2.0 platform model exploitation. As Jordan Murray, cofounder of Supp, explains,[21] "We believe the gig economy is essential to the future work mix, but the corporate model used by Uber and Airbnb leads to poor outcomes for workers and users who are the backbone of these platforms. Through decentralisation we plan to make a platform where there is no distinction between network owners and network participants. The concept is called a 'decentralised autonomous co-operative' and our MVP has already created more than AUS$170,000 in shifts for its members in Melbourne."

Supp connects workers to hirers just like any gig worker marketplace. Payment of wages is made in fiat currency; however, a token is automatically deposited into a worker's "tip jar" upon completion of a shift. Token economics running as smart contracts on Supp's blockchain not only ensure that the added value of participation returns to the participants but also reflect the shared principles of the community and thus deliver many positive externalities aside from pure commerce. Supp is a "decentralised autonomous co-operative" because of its democratic governance. It members own tokens that give them rights to vote ("one token one vote"). Supp's governance model has in-built provisions written in its smart contracts that encourage good behavior. For instance, a worker stakes a token on a shift upon accepting, and in the event of no-show the token is forfeited to the hirer. Tokens unlock governance rights and have increasing utility over time. As their value increases, tokens can also be exchanged for fiat money over cryptocurrency exchanges.

A particularly interesting characteristic of Supp's use of blockchain is how it handles user data. In web 2.0 platforms, user data are siloed and owned by the platform. Arguably, one of the reasons why these platforms are so exploitative is because of personal data "lock-in": workers cannot move their data to another platform, and indeed the cost of transition becomes higher over time. Rent-extraction machines on web 2.0 succeed because workers are vested in the platform and have little choice but to agree to onerous terms and conditions. By contrast, web 3.0 cryptoplatforms such as Supp offer self-sovereign identity and full data ownership rights. Supp users own their data on the blockchain and can take their data with them anytime they wish and join another platform. These data include their personal data as well as their reputation data. By pioneering data self-sovereignty in practice Sapp is pointing to a future where citizens are not powerless spectators but active participants in shaping the destiny of their countries and nations.

DATA SELF-SOVEREIGNTY AND DATA TRUSTS

Data property rights are an essential element for solving the wealth asymmetries of the Fourth Industrial Revolution.[22] In the AI economy algorithms are the pipes that deliver business value, and data are the fuel, the "new oil." There are several categories of data that AI systems need in order to train. What concerns us here are data generated by citizens. There are at least two classes of citizen data. The first comprises those that are strictly personal, such as data about one's age, sex, body measurements, achievements and failures, preferences and dislikes, health and employment records, and so on—let's call this class "personal data."[23] A second class of citizen data comes from social interactions and economic transactions, for example, data created each moment we interact via a digital channel with other people, groups of people, companies, institutions, courts, governments, and so on—let's refer to this class as "social data." Not all data have equal value, and most data only have value when they aggregate with other data. For example, the RAND Corporation calculated that in order to get accurate AI models for self-driving cars, there is a need to collect data from 500 billion to 1 trillion miles driven.[24] Unless individual drivers start sharing their driving data, building deployable models for self-driving cars may prove to be an impossible task for any company. Finally, there are data over which citizens must have complete control with respect to who has access,[25] and other data over which they must have partial control or none at all, for example, their medical or criminal records. Today, most citizen data are owned either by governments or by private corporations. This puts citizens at an economic disadvantage since they have no stake in the economic value that their personal and social data generate.

Regulation of personal data is currently restricted to privacy only, and is mostly centralized. The European Union's GDPR is an example of centralized governance of personal citizen data. Although the regulation is aiming to protect user privacy, it falls short from granting full property rights of the data to their rightful owners. Arguably, any centralized arbitration over data rights is subject to the intrinsic problems of representational governance: power asymmetries will always result in the strongest and the richest having more influence over the final result of arbitration compared with the weakest and the poorest. The principal-agent problem will emerge every time a centralized authority makes decisions on behalf of the many. Regardless of how "enlightened" the EU leadership may be, or how sincerely they may care for the data rights of European citizens, web 2.0 technologies will always favor the owners of the platforms more than the data-producing users.

One way of solving the problem of redistributing more widely the value created by our own personal data is to set up "data trusts." This is an idea pioneered by the *Open Data Institute* in the United Kingdom.[26] Data trusts can be mutual organizations with fiduciary responsibilities that act as stewards of citizen data. Citizens sign up to data trusts and pool their data together in order to make them available to businesses, governments, or other organizations that may be interested in using them. Data trusts negotiate with those interested organizations and offer access to the data in exchange for fees or a share of Intellectual Property (IP) rights derived by their use in training and powering AI algorithms and applications. Data trusts can be organized in various ways—for example, similarly to traditional pension trusts where a board of trustees has fiduciary responsibility for how the trust invests and distributes wealth. Such representational forms of governance do suffer, as discussed, from the principal-agent problem, which could be overcome by implementing data trusts on a blockchain and using cryptoeconomics to democratize its governance. Decentralized data trusts based on cryptogovernance models can be set up by cities, NGOs, private companies, worker unions, or grassroots citizen movements. They solve the principal-agent problem of democratic governance while accelerating innovation in AI by unlocking economic value in citizen data. They could also provide a framework to protect property rights for citizen data and thus contribute to reducing wealth asymmetries in the AI economy of the future. Finally, they could provide a defense against both corporate and state surveillance. As long as citizens are in control of their data, their privacy is more secure.

DECENTRALIZING AI

Today, several problems hinder the wider adoption of AI by small business and society at large. They range from lack of a clear strategy and shortage of specialized skills—such as adequate numbers of AI engineers and data scientists—to shortage of good and varied data sets. Eager to regulate, governments are adding onerous compliance costs on top of all that, making it even harder for small businesses that aspire to ride the growth wave of AI. Thus, regulation and scarcity of skills and data combine to favor the big tech giants that can use their dominant position and economics of scale to advance AI ahead of everyone else. In other words, the AI economy is not a level playing field but is massively influenced by an oligopoly of players. Is there a way to change the rules of the game?

The open-source movement has shown how peer-to-peer networks can accelerate innovation by creating communities that share values and knowledge, as well as code. Cryptoeconomics can take this proven paradigm to a whole new level by adding economic incentives and democratic governance. By putting those two things together, we can imagine a "decentralized AI marketplace," an exchange for AI tools, data, and applications built using blockchain technology, where network effects and cryptoeconomics accelerate innovation and lower costs, while significantly lowering the barriers for participating in the AI economy. To deliver such a marketplace, one needs three principal building blocks. The first block is a data marketplace, where anyone can provide his or her data securely in an open data exchange. Ocean Protocol[27] is one of the most prominent innovators in this space. By creating a cryptonetwork for data, they are delivering a way for citizens, communities, and business to share data on a marketplace securely and transparently while keeping ownership. Data trusts could also be major contributors in the supply side of such decentralized data marketplaces.

The second building block for a decentralized AI marketplace is decentralizing cloud computing so that AI innovators can train models in a cost-efficient way, without lock-in to the cloud of big tech providers. Currently, just four centralized cloud computing providers—Microsoft, Amazon, IBM, and Google—own more than 70% of the cloud computing market.[28] This oligopoly prevents prices from dropping to a level where anyone, anywhere in the world, can become an innovator and create economic value. The cost for access to computing will become an even more critical factor that will decide the wealth asymmetries of the future. As we enter the Fourth Industrial Revolution, deep learning networks will be one of the most proliferative technologies: they will power image and language recognition systems as well as a host of other applications embedded across business systems and infrastructures.[29] These technologies are computationally intensive, and this is one reason why the current cloud computing oligopoly is at odds with the interest of society at large: it raises cost-to-access barriers and skews wealth creation toward capital owners and against ideas creators and data owners. The UK-based start-up DADI[30] is building a decentralized cloud computing marketplace, with a vision to provide a viable and preferable alternative to the current cloud computing oligopoly. DADI uses blockchain to create a peer-to-peer network where anyone can connect his or her computing device and earn income by providing spare computational capacity for rent. Decentralized cloud computing services, such as DADI, are challenging the status quo.

The third building block is a decentralized marketplace for AI models, applications, and tools; that includes cost-efficient frameworks where innovators can create, assemble, and market their ideas. A decentralized AI marketplace ecosystem that interconnects data, computing, and development frameworks is a hub for fast innovation and democratic economics.

Here's how the whole thing would work: developers solve for real business problems and deliver innovative AI applications to businesses, organizations, and communities. They do so by collaborating with end customers over the AI marketplace in order to define the problem that needs solving. They then deliver the solution by accessing, combining, and evolving more elementary AI tools and models, which are supplied by the wider AI developer community on the platform. Since data are the fuel for successful AI applications, these AI innovators access unique data sets that are supplied through the decentralized data exchanges and data trusts. Training and deploying the models takes place using the decentralized computing infrastructure, which lowers the cost of innovation and systems use. Using cryptoeconomics, tokenized rewards flow across the wider ecosystem, and each participant on the supply side (AI tools and applications, data, computing, ideas, etc.) is rewarded according to the participant's contribution. This creates powerful incentives for all participants. Artificial intelligence innovators can monetize their work, and businesses, communities, and individuals can turn their business data into profitable assets. Customers get what they need at a very low price. And there is no lock-in to a centralized cloud provider. Everybody wins.

THE BIG TRANSFORMATION

Artificial intelligence is a general-purpose technology that will be the main driving force of the innovation and wealth creation that will take place in the Fourth Industrial Revolution. In combination with blockchain-powered cryptoplatforms, the dividends from the emerging AI economy can be distributed more fairly and widely and thus help reduce the wealth and income asymmetries that are currently fueling popular mistrust in free markets and the institutions of liberal democracy. In the years to come, many other technologies will enter the "plateau of productivity" with profound impacts on our world. One could imagine, for example, the tremendous impact of decentralizing energy production using ultraefficient solar panels at every home, and what that would mean for localizing energy distribution grids and reducing transmission costs and environmental degradation.[31] Or, the significance of decentralizing manufacturing using 3D-printing technologies;[32] or, perhaps, the revolutionizing of food production using a combination of urban farming and genetic engineering so that everyone can produce most of their food closer to home.[33] In all those cases citizens will become producers and consumers of their fundamental needs at the same time. We have witnessed a similar transformation during the first phase of the information revolution, with the advent of social media, whereby citizens became both the producers as well as the consumers of content. Web 3.0 technologies have the potential to extend this transformation by including just about everything and can thus create new markets of millions of "prosumers" where political power is inherently decentralized.

In such a truly interconnected peer-to-peer society citizens can own tokens across many web 3.0 platforms. Smart financial systems and exchanges can help them manage and invest or trade their tokens, so that they can absorb troughs when income might be scarce, as well as pay for the provision of essential services in health care, insurance, education, and retirement. Interconnected ecosystems of democratically governed cryptoplatforms that power AI systems can reinvent the "sharing economy" by distributing the creation of new wealth among their token holders, while rewarding fairly those who contribute more to the collective welfare. The digital tokenization of these new peer-to-peer markets could thus reform capitalism and make it work for the many and not just the few.[34]

Moreover, these new peer-to-peer markets can transform how we think of ourselves as workers and active members of society. In traditional capitalism workers are a potential liability, for they must produce more than they consume in order to have any value. In a tokenized web 3.0 economy workers become value-producing assets by virtue of their existence, as value always feeds back to them from their individual inputs. Collectively, this would mean, in effect, the true realization of an autopoietic cybernetic system incorporating humans and their technologies at a societal scale. For, whether it will be our data, or our other intellectual or physical contributions, we would be providing via our joining, participation, and interactions with the various peer-to-peer cryptoplatforms the raw energy that could power the engines of sustainable and resilient economic well-being in the future.

10

SCALING A POLIS

The broad outline of a Cyber Republic should have emerged by now. Many details are still missing, but the purpose of this book has been to provide a first iteration, an imperfect yet adequate paper prototype to trigger a broader discussion for an evolved system of liberal government fit for the Fourth Industrial Revolution. I have argued that such evolution is necessary to address the sense of unfairness and inequality that has gripped millions of citizens following the global financial crisis and the Great Recession, to strengthen our democratic institutions, and to prepare our societies for the impact of AI on work and wealth creation. We therefore need to find ways to clean up politics by broadening democratic participation, but also to reform free markets. With this purpose in mind, I have focused on three specific problem areas. The first is how to extend citizen rights so that we become more directly involved, and personally responsible, for policy making. I have argued in favor of using a citizen assembly model in order to complement representational democracy, and I have shown how conversational AI systems based on cybernetics can be used to solve the problem of knowledge asymmetry between experts and nonexperts. Second, I proposed to reconnect AI to its cybernetic roots, so we may avoid the existential risk of an artificial superintelligence having goals that are different from or conflict with those of humans, but also in order to unlock new economic and social value from new types of human-machine collaboration. Third, I tried to present an overview of how we could leverage web 3.0 cryptoplatforms and cryptogovernance, combined with data and AI, in order to deliver new, peer-to-peer markets that can reform capitalism, solve problems of wealth and income asymmetry, compensate for loss of income due to automation, and ultimately democratize the AI economy of the twenty-first century.

In this chapter, I would like to add some more detail in the Cyber Republic paper prototype by looking into how we can use cryptogovernance to transform aspects of government. For this, I will again narrow my focus to two specific areas of government responsibility. The more important of the two is, perhaps, the management of the commons, of things like public infrastructure, utilities, and natural resources. Governments have traditionally managed the commons by enacting protection laws and by applying centralized oversight and regulation of the oligopolistic markets that exploit the commons. I will therefore look into the possibility of an alternative, decentralized model to manage the commons that might be more efficient and more effective. This model will be based on the economic ideas proposed by Nobel laureate Elinor Ostrom (1933–2012), and I will argue that it could be realized using the same cryptogovernance approach that was discussed in the case of peer-to-peer markets.

But first, I will look into how local city governments can become platforms of technological and social innovation by creating data trusts for citizen data, and I will discuss how this approach could further invigorate citizen trust in liberal institutions. Cities are the places that provide the greatest opportunity to reinvent our liberal system of government. They are the veritable engines of economic growth. Over 4 billion people live in cities today, more than half of the global population, generating more than 80% of the global GDP.[1] According to the UN, and as urbanization continues, by 2050 the number of city dwellers will rise to 68% of the world population.[2] The economic potential of cities is enormous. In 2010 scientists from the Santa Fe Institute showed that every time the population of a city doubles, citizen productivity per capita goes up by 130%.[3] In fact, not only does productivity increase but innovation does too. The bustling chaos of city life is a cauldron for fermenting new ideas, inventions, experiments, and discoveries. The very word "civilization" comes from the Latin word *civitas*, which means "city."[4] But there is a caveat. Not all cities can deliver economic prosperity. Many megacities in the developing world are falling behind despite the millions that move there. At the same time six hundred cities—the City 600 Index compiled by McKinsey—are projected to generate more than 60% of global growth to 2025.[5] What separate the City 600 Index from the rest are the high levels of security, the wide availability of public spaces, and the excellent communications and transport facilities. In other words, for cities to become powerhouses of economic growth, they must provide opportunities for interactions and connectivity.[6] City dwellers need to meet people outside their inner circle, engage in accidental conversations, interact with novel ideas, explore art shows, and listen to talks and debates. Without a rich and efficient web of connectivity, cities are mere amalgamations of small isolated villages clustered together. So how can we maximize connectivity in cities? Data, AI, smart sensors, and the Internet of Things are some of the tools that we can use to take connectivity in cities to a whole new level and transform cities into "smart cities." But technology without proper governance can easily lead to undesirable outcomes. A typical example of this has been the smart city that Google aspired to create in Toronto, Canada.

DATA TRUSTS FOR SMART CITIES

In 2017 Sidewalk Labs,[7] a sister company of Google, earmarked 800 acres of disused land in Quayside, Toronto, to build a smart city from the ground up that would include ultrafast Wi-Fi, driverless cars, innovative ways to collect rubbish using subterranean tunnels, and a dense network of sensors. But the company's vision of a sensor-laden neighborhood relaying data to a private company quickly run into trouble, triggering much controversy among Toronto citizens and civic groups. The Canadian Civil Liberties Association sued the Canadian government over the plans, arguing that it was "inappropriate" for a firm like Google to design privacy policies to govern city neighborhoods.[8] The privacy advisor of Sidewalk Labs ended up resigning a year into planning, writing in her letter of resignation, "I imagined us creating a smart city of privacy, as opposed to a smart city of surveillance."[9] Many other voices protested against "living in a lab" run by a private firm. The protests revealed how the business model of web 2.0 platforms is a mismatch for civic life. Google's winning formula for digital has been to harvest user data in exchange for free services, such as search or email. Transferring this formula to the physical world of a city failed to account for the difference between digital and civic experience. Cities are complex ecosystems with their own internal dynamics.[10] We may not think twice when asked on a web page to give consent to cookies, but it is a completely different game to know that our moves are watched over by cameras and that the data are fed to a private company that is largely unaccountable for what it might use those data for. Google's strategy in Toronto backfired because citizens are nowadays more aware than ever of the perils of surveillance and loss of privacy. Following the backlash and the protests, Sidewalk Labs recoiled from its original ambitions to harvest and own every piece of citizen data and called for the establishment of an independent civic data trust that would be responsible for data governance.[11] By doing so, it conceded to the emerging trend that citizens of smart cities are not willing to share their personal data with private companies directly but would rather trust an intermediary organization with fiduciary responsibilities and, preferably, democratic governance.[12]

Meanwhile in England, the Open Data Institute collaborated with the Royal Borough of Greenwich in London in one of the first pilots in the world to design, build, and test a civic data trust.[13] The pilot focused on data about electric vehicle parking sensors and data collected by heating sensors in residential housing and aimed to develop best practices for balancing, on the one hand, citizens wanting to ensure that their data are not used to cause harm and that their privacy is protected and, on the other hand, the benefit of giving access to innovators who would like to use these data in order to develop new products and services that will improve citizens' lives. Although it is still early days in the establishment and launching of civic data trusts, and there is still much to explore and learn, the appetite for using them is growing. The potential for transforming local and national governments is enormous. Citizen data are an essential part of a city's and a nation's infrastructure, especially as governments undergo digital transformation and try to reinvent themselves as key players in facilitating and accelerating, rather than impeding, the digital economy. Therefore, it is vital to having a way to secure and protect citizen data, while giving access to that data to drive economic growth. Civic data trusts can provide the required institutional layer, as well as the data architecture, to do this successfully. The London pilot demonstrated how a data trust could evolve further by incorporating more data sets and thus provide opportunities for exponential innovation.

Consider, for example, the interdependencies between traffic lights and energy; waste and transport; and communications and energy. The London pilot showed that traffic-monitoring sensors for a congestion-management application could also accommodate environmental and other sensors and thus deliver open hardware and communications layers with a high potential for reuse. In a future iteration, a smart city might import additional data from multiple other sources—for instance, social media data to improve its sensing capabilities and as an information feedback mechanism to alert car drivers, cyclists, and pedestrians. Creating such an infrastructure of data and sensors would allow deep integrations into city planning tools for housing, green areas, and siting of hospitals and old-age care centers that could improve citizen health and safeguard wellness. Such an infrastructure would also allow the expansion of a smart city's applications with additional capabilities, such as AI systems, policy controllers to enforce privacy and security, and tokenization capabilities through a public blockchain that can enable citizen data monetization.[14] A civic data trust for a smart city could thus be a platform for building applications that reinvent administration processes and vastly improve the experience of citizens transacting with government departments as well as business. So imagine several apps that can help you file your taxes, or start a company, or book an appointment with a hospital. Citizens would access those apps via a decentralized "app store" and select and pay for the ones that they like most. By combining market economics and data trusts, cities can revolutionize how their governments serve their citizens. This approach to protecting, enriching, and sharing citizen data can also scale from the local to the national level.

UNLEASHING GROWTH BY ENCRYPTING CITIZEN ID

For example, one may imagine a national data trust that delivers encrypted "citizen IDs" that include every record about a citizen, including heath records, education, work experience, skills, financial information, and so on. Each citizen ID data record could be kept on a decentralized, blockchain-powered data trust that is cogoverned by elected representatives plus citizen representatives whose records are on the chain. Given the properties of cryptonetworks, discussed earlier, access to citizen data will be encrypted but also allowed on a get-only-what-you-need basis. For instance, if a bank needs to check credit history, it will not get access to a citizen's full financial data but will get only a green, amber, or red light reflecting the citizen's credit risk. The same service could extend to immigrants, to help them create and validate new identities and records as they settle and integrate in their new homes.

Such types of data trusts are already being explored in Africa, where one of the biggest problems is financial exclusion of millions of unbanked citizens. The Kiva Protocol, pioneered by the Sierra Leone government and Kiva, the Silicon Valley microloan company, is a blockchain-based biometric system that links a person's thumbprint with his or her identity.[15] In Sierra Leone, where per capita GDP is about US$500 a year, just 20% of the 5.1-million adult population have bank accounts. By creating a universal credit bureau for its citizens, Sierra Leone is hoping to spur lending by banks reluctant to loan to people without credit histories. But the country's ambitions are greater than that. Having a secure and trusted citizen ID system can offer additional benefits that include allowing the government to reach more people with its services, cutting costs for mobile operators and start-ups, and bringing thousands of small businesses into the formal economy. As the country's chief innovation officer said in an interview, "From the individual, to the start-up to the government to business, the proof of ID becomes instantaneous, meaning more access to services for Sierra Leoneans. This is a great step that a small country is taking."[16]

Sierra Leone is showing the way to solving one of the biggest problems in the world today. A recent report by the World Bank estimates that roughly 1 billion people worldwide, the vast majority of whom live in sub-Saharan Africa and South Asia, lack basic ID credentials.[17] McKinsey estimates are much greater, pointing to 3.4 billion people who have some sort of ID but limited ability to use it in the digital world.[18] Those "invisible billions" include children without registered birth certificates, women, disabled and rural inhabitants, and refugees. As the World Bank report highlighted, all these people find it difficult to access health care and social services, enroll in school, open a bank account, obtain a mobile phone, get a job, vote in an election, or register a business.[19] Inclusive and trusted digital ID systems can unleash the economic growth potential of billions and generate new markets and also reduce fraud and leakage in the delivery of public services. Data trusts, being systems and architectures for data governance built with privacy and security by design, can deliver the required solution to this huge, global problem of trusted IDs and thus help enhance democratic systems of government through inclusion and transparency.

CRYPTOGOVERNANCE FOR THE COMMONS

How decisions are made, and by whom, in sharing resources that belong to no one and yet to everyone at the same time has been one of the biggest problems in political governance. These "commons" may range from basic infrastructure and natural resources to public spaces and buildings, or less tangible things like access to knowledge or data. Sharing the commons is neither easy nor straightforward. The American ecologist and philosopher Garrett Hardin (1915–2003) coined the term "tragedy of the commons" to underscore how individuals acting on self-interest can destroy a shared resource. A well-documented example of failure to manage a shared resource is the Grand Banks Fisheries off the coast of Newfoundland in Canada. The bountiful supply of cod that sustained human communities for centuries was challenged in the 1970s when improved fishing technologies allowed much larger catches. By the 1990s cod populations had collapsed, and as result everyone in the fishing industry suffered.[20] In a centralized world the tragedy of the commons is usually addressed by enforcing top-down regulation by a central government or an international agreement between central governments. But is there another, more efficient and more democratic, way to manage the commons? One that may be a better fit for an evolved and decentralized system of liberal government, as Cyber Republic advocates? The answer has been given to us by the first woman to win the Nobel Prize in economics, Elinor Ostrom.

Ostrom's research on what she termed "common pool resources" (CPR) demonstrated that, within communities, rules and institutions can emerge from the bottom up to ensure the sustainable and shared management of resources in an economically efficient way. There are several historical examples that demonstrate how a bottom-up, self-organized system of rules is more economically efficient than the centrally planned rules and regulations of a Leviathan state. The Enclosure Acts promulgated in England in the eighteenth and nineteenth centuries privatized lands that were hitherto "common," used for grazing livestock, hunting, or cultivating.[21] Privatization of the commons led to a more intensive and productive cultivation of the land but had many unintended and dramatic consequences. English shepherds and farmers depended on freely using the commons; with access now limited or restricted they were forced to move their families from the countryside to the cities, in what economist Karl Polanyi defined as the "great transformation" of English society.[22] This new, low-cost labor that gathered in cities was key in exploiting Watt's technological breakthrough—the steam engine—and fueling the First Industrial Revolution.[23] But at what a terrible human cost! Families had to live in squalid conditions, their physical and mental health severely impacted. It took a century for the welfare of workers to start improving, with new labor laws and a welfare state.

Ostrom, writing in her seminal book *Governing the Commons*,[24] noted that at the same time England was enacting the Enclosure Acts, a substantially different definition of the law on land was made in Törbel, Switzerland. There, the management of land was assigned not to a private or a government entity but to the community of people that used it. This was common and stable practice at Törbel for centuries and clearly demonstrated that a "third way" was more efficient and effective than either top-down government regulation or privatization. Besides Törbel, Ostrom showed examples of common lands managed by communities in the Japanese villages of Hirano and Nagaike, the *huerta* irrigation mechanism between Valencia, Murcia, and Alicante in Spain, and the *zanjera* irrigation community in the Philippines.[25] The property in the form of vicinal neighborhoods, typical of regions of Italy like Emilia and the Belluno, the Ticino in Switzerland, as well as the eighteenth-century Ampelakia commonwealth in central Greece,[26] are also examples of collective, bottom-up institutions that have managed commons efficiently. Modern examples are the open-source movement in software and, of course, Wikipedia, wherein the commons is knowledge.

Ostrom identified eight design principles of stable, self-organized, local CPR management.[27] The first condition is the clarity of the law, but in addition to being clear rules must also be localized and shared by the community. Another essential element of self-government is the establishment of collective and democratic decision-making that involves all users of the resource. Mechanisms of conflict resolution must be local and public and must be accessible to all individuals of the community. Furthermore, control of the resource must be exercised by the users themselves.

Ostrom's prescriptions for self-regulated commons could be implemented in the constitutional layer of a cryptoplatform (see figure 9.2). Once these prescriptions have been deliberated in a citizen assembly, they could be coded into smart contracts on a distributed ledger that would control the management, or operating system layer, of the commons. A value-adding layer and a supervisory layer would then follow. This layered approach effectively creates a decentralized system for the governance of the commons where principles of cryptoeconomics may also apply. Thus, the cryptogovernance platform could reward those who contribute to consensus, that is, in the enforcement of the governance rules of the decentralized, blockchain protocol.[28] As an example of this taking place already, the firm Benben in Accra is developing land-title registries for Ghana using blockchain;[29] while the countries of Georgia and Honduras are also doing the same.[30] This decentralized, blockchain-powered approach protects local populations more effectively from eviction by powerful corporations or corrupt governments. Smart contracts could be used to also underwrite the property rights of indigenous or local populations to natural resources in their area, including mines and fisheries. For example, fish might be traded on a platform only if local communities have approved harvest quotas by a democratic process.[31]

Cryptogovernance of the commons could unlock economic value that remains currently untapped because of the high compliance costs and inefficiencies that usually accompany any top-down regulation, or because of oligopolistic models of restricted privatization, or both. We have already seen how decentralized civic data trusts can accelerate innovation in smart cities by creating open platforms of shared data. We can only imagine the potential for economic growth that liberal democracies can unleash by democratizing the governance of the commons.

EPILOGUE: USING *CYBER REPUBLIC*

The purpose of this book was to analyze aspects of the problems with our politics, our technology, and our economy in the context of the Fourth Industrial Revolution and to propose ideas for debate and discussion. Some of those ideas are in production, others have delivered pilots or proofs of concept, and others need more elaboration and, most importantly, experimentation. Data trusts are a novel concept that has to be adopted by a critical number of pioneering smart cities and organizations in order to become the mainstream paradigm. Their use in the COVID-19 pandemic could have resolved the dilemma of protecting data privacy while permitting the digital innovation that needed that data in order to help countries return to normality. But the big prize from data trusts is, of course, the institution and protection of property rights for citizen data, an idea that needs more discussion and exploration, as well as input from legal and economic scholars. Cybernetic AI systems are a proposal that goes against the current trend, and investment, that is betting on traditional AI systems that are decoupled from humans. Web 3.0 cryptoplatforms are in their infancy. Whether they replace web 2.0 platforms as the dominant business model of the Fourth Industrial Revolution is uncertain, and—like the development of cybernetic AI—contingent on how much investment, talent, and sense of purpose will be diverted toward more democratic systems of wealth creation and corporate governance.

Much depends on the extent to which liberal democracies will encourage the development of open, peer-to-peer markets based on web 3.0 technologies through the appropriate legislation and regulation. Given, however, the growing realization that only a global cryptocurrency could provide the necessary stability to the global banking system,[1] we could be hopeful that this would be the general direction of travel for many governments and regulators in the 2020s. Should such a global cryptocurrency ever be launched and adopted by the major economies, it will pave the way to an explosion of web 3.0 cryptoplatforms and peer-to-peer markets wherein tokens could be valued against this stable, global cryptocurrency.

Citizen assemblies need to become a new liberal institution, embedded in the policy decision making at every level of government. The COVID-19 pandemic demonstrated how important it is for democracies to get citizen consent to draconian measures that directly impact their lives. The pandemic also illustrated the need for democratic societies to find balanced solutions and make hard choices for the restart of their national economies. Leaving those decisions to the experts and the politicians, without directly engaging with citizens, exposed extremely difficult and costly dilemmas. How many people should lose their jobs, and how many businesses should fail, before lifting the lockdowns? How should we allow the gradual return of people to work without introducing massive surveillance, socioeconomic exclusion, and division? How can we evaluate scientific knowledge and technological innovations that are imperfect, yet urgently needed? How can we balance the rights of the vulnerable old with the rights of the less-vulnerable young? Dilemmas such as these are not scientific but moral and political. Ordinary citizens ought to have a voice in reaching a consensus on the optimal solutions and strategies, and take the risk, as well as the responsibility, for implementing those solutions without the need for top-down enforcement or coercion.

The concept of cryptogovernance based on Elinor Ostrom's design principles for CPR institutions is a speculative proposal that requires testing. The potential of Ostrom's thinking with respect to transforming institutions to widen participation, promote diversity, and favor cooperation over competition is enormous but as yet largely unexploited. There is a need for her work to be further popularized so that it inspires young scientists, technologists, and entrepreneurs in experimenting more with alternative business models and, importantly, with new ways of managing the commons. The twenty-first century will require that human societies find consensus to tackle global challenges together. By the year 2050 there will be nearly 10 billion humans living on the planet.[2] As more people come out of poverty, and assuming that average human prosperity keeps increasing, planetary resources will stretch to their limit. Technology can provide temporary solutions for food insecurity, energy, provision of fresh water, and a slowing down of environmental degradation. However, how we apply those technologies to protect and manage the planetary commons will be critical to our survival. Current centralized institutions of global governance seem ineffective and outdated when it comes to dealing with such huge challenges. New institutions for global governance are needed that decentralize decisions and actions so that there is greater citizen involvement in decision making and policy making at a local level; Ostrom's ideas could provide ample inspiration for imagining such institutions.

Cryptoeconomics needs to expand beyond the design of smart contracts and explore how web 3.0 technologies could enable new business models for a circular economy. For instance, by having a community of consumers using a cryptoplatform to share the use of the same products, we can drastically decrease the energy expended for manufacturing, while at the same time increasing revenues for producers. Combining circular economic models and cryptoplatforms, we could provide clear economic incentives to ordinary citizens for sustainable living without loss in the quality of life.

As we design new digital systems for production and governance, and as we evolve our political institutions to allow for more direct citizen participation, it is important to take into account scientific discoveries that are shedding new light on human empathy and morality. The neurobiology of conscience and moral thinking should inform the cybernetic designs of the future, and that is why designers and engineers must become acquainted with interdisciplinary work in how evolution is shaping moral behavior by groundbreaking academic thinkers such as Nicholas Christakis,[3] Bret Weinstein,[4] and Patricia Churchland,[5] to name but a few. We need more such interdisciplinary work in order to connect the dots of evolutionary biology, sociology, politics, philosophy, ethics, history, and engineering. For example, in a future of material abundance, when most people would be free to choose how much they work and when, what are we going to do with our freedom? How far can we go by replacing the dignity of having a steady job with the dignity of being an active citizen in a participatory democracy? What can we learn from the state of humanity before the agricultural revolution, or from present-day preindustrial communities, to inspire productive and meaningful ways of living in a largely automated future? What can we learn by reminding ourselves of ancient and contemporary philosophical teachings that explored a "good and considered" life, so we can incorporate them in the context of a technologically advanced civilization? How can we use those teachings to preserve and protect natural habitats, explore new worlds, and save humanity from self-destruction? The current state of siloed academic discourse is unhelpful, and universities must do more to break down functional barriers and sponsor more horizontal research. More dialogue is needed between technologists, politicians, and theorists. *Cyber Republic* is an attempt to provide ideas and a framework for such a discourse but also for direct action.

For the struggle against authoritarianism and absolutism must go on and intensify not only in academic circles, books, and intellectual debates but also in the way we run our businesses, educate and inspire the next generation of citizens, and organize collectively in our workplaces, our industries, our professions, and our cities. We must not let down our guard and give in to the sirens who are using the COVID-19 pandemic as a pretext to reduce civil liberties and adopt a surveillance state with little or no accountability to citizens and communities. We should take hope and inspiration from the millions who filled the streets of Hong Kong and raised their voices in defense of liberty and the rule of law. And we should borrow some of the great courage of 17-year-old journalism student Olya Misik, who, during the anti-Putin demonstrations in the summer of 2019, read articles from the Russian constitution to the heavily armed riot police deployed to break up the rallies.[6] As long as there are people willing to fight tyranny regardless of personal cost, the future is open and full of possibilities. We can claim it. We can still dream of a better world.

ACKNOWLEDGMENTS

Many have contributed over many years to the development of the ideas presented in this book. The Eugenides Foundation in Athens and the King Baudouin Foundation in Brussels trusted me with the role of facilitator in the Meeting of Minds project, which started everything. My role as external relations officer for the European Bioinformatics Institute, as well as my stint as trainee and contractor at the European Commission, gave me insights into the inner workings of European political integration and politics—and I would like to thank the many people in both institutions who generously opened doors for me. I want to thank my colleagues at Willis Towers Watson for their friendship and support, and particularly Ravin Jesuthasan, Tracey Malcolm, Richard Veal, Suzanne McAndrew, and Karen O'Leonard for giving me the opportunity to explore, learn, debate, and apply many new ideas on how businesses could transform as a result of AI and big data. It is important to note here that the views expressed and presented in this book are mine and do not necessarily reflect the views of Willis Towers Watson. And, of course, my warmest thanks to my clients, too, from across a wide range of industries, all of whom are undergoing so much change and disruption because of AI. Working with such great clients has reinforced my conviction that business is where changing the world for the better begins. My most sincere thanks also go to Paul Pangaro at Carnegie Mellon University for introducing me to

Gordon Pask; Geoff Goodell at University College London for his insights into the application of blockchain and smart contracts in financial services; Vince Kuraitis for his permission to reproduce the comparison table between web 2.0 and web 3.0 platforms; Andrew Konya and Gary Ellis at Remesh for brainstorming on the use of AI to scale deliberative democracy; Wally Trenholm and Paul Copping at Sightline Innovation for sharing their experience on data trusts for smart cities; Amir Baradaran at Columbia University for inviting me to give the keynote at the Another AI in Art summit in New York, and our lengthy discussion on inclusion and decolonization of AI; Catherine Havasi at MIT Media Lab for her unique insights on natural language processing and AI ethics; and Josh Sutton at Agorai for generously sharing his vision for AI with me. Last, but certainly not least, I would like to extend my thanks to two people who were instrumental in the publication of *Cyber Republic*: Alexander Cochran, my amazing agent at Conville & Walsh, and Marie Lufkin Lee, my wonderful editor at the MIT Press, who supported and guided me throughout the writing of this book.

NOTES

FOREWORD

1. Don Tapscott, "A Declaration of Interdependence: Toward a New Social Contract for the Digital Economy," Blockchain Research Institute, rev. January 14, 2019, https://www.blockchainresearchinstitute.org/socialcontract.

2. "2020 Edelman Trust Barometer," Daniel J. Edelman Holdings Inc., January 19, 2020. https://www.edelman.com/trustbarometer.

INTRODUCTION

1. Troika (in the context of the Greek crisis): the European Union, the European Central Bank, and the International Monetary Fund.

2. Oana Lungescu, "Greece Should Sell Islands to Cut Debt," BBC News, March 4, 2010, http://news.bbc.co.uk/1/hi/8549793.stm.

3. Sebastiaan Faber and Bécquer Seguín, "Welcome to Sunny Barcelona, Where the Government Is Embracing Coops, Citizen Activism, and Solar Energy," *The Nation*, August 11, 2016, https://www.thenation.com/article/welcome-to-sunny-barcelona-where-the-government-is-embracing-coops-citizen-activism-and-solar-energy.

4. The Icelandic citizen assembly (the "2010 National Gathering") took into account suggestions put forward by 1,000 randomly selected citizens ages 18 to 89 years old. It should be noted that the constitutional proposals of the Icelandic citizen assembly were ultimately rejected by the national parliament. Richard Conner, "Iceland Votes for Citizen Assembly to Draft New Constitution," DW, November 28, 2010, https://www.dw.com/en/iceland-votes-for-citizen-assembly-to-draft-new-constitution/a-6274235-0.

5. Joseph Stiglitz, "How I Would Vote in the Greek Referendum," *The Guardian*, June 29, 2015.

6. Winston Churchill, quoted in Bartleby.com, at https://www.bartleby.com/73/417.html.

7. Rachel Donadio, "Macron and Salvini: Two Leaders, Two Competing Visions for Europe," *The Atlantic*, May 20, 2019, https://www.theatlantic.com/international/archive/2019/05/emmanuel-macron-matteo-salvini-europe/589753.

8. Klaus Schwab, *The Fourth Industrial Revolution* (New York: Crown, 2016).

9. The Second Industrial Revolution starts at the end of nineteenth century and through the first part of the twentieth century with rapid industrialization, mostly due to the invention of electrical power production and distribution. The Third Industrial Revolution started in the second part of the twentieth century with the invention of the computer; we are arguably still in that historical phase that saw the emergence of the Internet and the digital economy.

10. Given that, on average, labor accounts for a significant percentage of total production costs. For precise information on labor contribution and effectiveness, read the US Human Capital Effectiveness Report published periodically by PwC. For example, https://www.pwc.com/us/en/hr-management/publications/pwc-trends-in-hr-effectiveness-final.html.

11. "Difference Engine: Luddite Legacy," *The Economist*, November 4, 2011.

12. We see this clearly happening already with companies such as Amazon, Google, Facebook, and Apple amassing billions of dollars by virtually controlling the digital economy, where the "digital platform" is essentially a rentier's dream.

13. China started opening up its economy to global markets in the late 1970s under the leadership of Chairman Deng Xiaoping. However, there can be no doubt that the Communist Party can intervene at any time to replace, and sometime imprison, corporate bosses that fall out of the party line. In fact, under President Xi Jinping the Communist Party has increased its control over state-owned and private enterprises.

14. As reported in Forbes, a 2019 survey showed a 50% failure in 25% of enterprises deploying AI. Gil Press, "This Week in AI Stats: Up to 50% Failure Rate in 25% of Enterprises Deploying AI," Forbes.com, July 19, 2019, https://www.forbes.com/sites/gilpress/2019/07/19/this-week-in-ai-stats-up-to-50-failure-rate-in-25-of-enterprises-deploying-ai/#3dc95aa372ce.

15. Henry Mance, "Britain Has Had Enough of Experts, Says Gove," *Financial Times*, June 3, 2016, https://www.ft.com/content/3be49734-29cb-11e6-83e4-abc22d5d108c.

16. These values are also stipulated in the Liberal Manifesto, adopted by the 48th Congress of Liberal International. See "Oxford Manifesto 1997: The Liberal Agenda for the 21st Century," accessed via https://web.archive.org/web/20110207012341/http://www.liberal-international.org/editorial.asp?ia_id=537.

CHAPTER 1

1. Astra Taylor, *Democracy May Not Exist, but We'll Miss It When It's Gone* (New York: Metropolitan Books, 2019).

2. Agnes Heller, "Hungary: How Liberty Can Be Lost," *Social Research* 86, no. 1 (Spring 2019).

3. Timothy Garton Ash, "Europe Must Stop This Disgrace: Viktor Orbán Is Dismantling Democracy," *The Guardian*, June 20, 2019, https://www.theguardian.com/commentisfree/2019/jun/20/viktor-orban-democracy-hungary-eu-funding.

4. Dalibor Rohac, "Hungary and Poland Aren't Democratic: They're Authoritarian," *Foreign Policy*, February 5, 2018, https://foreignpolicy.com/2018/02/05/hungary-and-poland-arent-democratic-theyre-authoritarian.

5. Yascha Mounk, *The People vs. Democracy: Why Our Freedom Is in Danger and How to Save It* (Cambridge, MA: Harvard University Press, 2017).

6. Isaiah Berlin, "Two Concepts of Liberty," *Four Essays on Liberty* (Oxford: Oxford University Press, 1969), 118–172.

7. John Stuart Mill, *On Liberty* (1859). 2016 edition by Enhanced Media Publishing, Los Angeles, CA.

8. Jean-Jacques Rousseau, *The Social Contract* (1762). 1998 edition by Wordsworth Editions Ltd., Hertfordshire, England.

9. Walter Lippmann, *The Phantom Public* (New Brunswick, NJ: Transaction, 1925).

10. Full transcript of Hillary Clinton's comment: http://time.com/4486502/hillary-clinton-basket-of-deplorables-transcript.

11. Mounk, *The People vs. Democracy*.

12. The European Convention on Human Rights can be accessed at https://www.echr.coe.int/Documents/Convention_ENG.pdf.

13. Given that Italian governments change quickly and unpredictably, I refer here to the coalition government of 2018–2019 between Five Star and the Northern League.

14. As the far-right conspiracy theory social media platform QAnon promotes.

15. Rasmussen Global and Dalia Research, "Global Perceptions of Democracy," June 22, 2018, https://www.allianceofdemocracies.org/wp-content/uploads/2018/06/Democracy-Perception-Index-2018-1.pdf.

16. Jaspers quoted in Benas Gerziunas, "Citizens Disillusioned with Democracy: Poll," Politico, June 21, 2018, https://www.politico.eu/article/democracy-europe-citizens-disillusioned-poll.

17. Francis Fukuyama, *The End of History and the Last Man* (New York: Free Press, 1992).

18. Data accessed March 2018, https://freedomhouse.org/report/freedom-world/freedom-world-2018.

19. Richard Wike, Katie Simmons, Bruce Stokes, and Janell Fetterolf, "Globally, Broad Support for Representative and Direct Democracy," Pew Research Center, October 16, 2017, http://www.pewglobal.org/2017/10/16/globally-broad-support-for-representative-and-direct-democracy.

20. Wike et al., "Globally, Broad Support."

21. Rupert Neate, "Richest 1% Own Half the World's Wealth, Study Finds," *The Guardian*, November 14, 2017, https://www.theguardian.com/inequality/2017/nov/14/worlds-richest-wealth-credit-suisse.

22. Oxfam's 2017 report "An Economy for the 99%," accessed via https://www.oxfam.org/en/research/economy-99.

23. Jill Treanor, "Half of the World's Wealth Now in Hands of 1% of Population—Report," *The Guardian*, October 13, 2015, https://www.theguardian.com/money/2015/oct/13/half-world-wealth-in-hands-population-inequality-report.

24. World Economic Forum, "The Global Risks Report 2017," accessed via http://www3.weforum.org/docs/GRR17_Report_web.pdf.

25. As a demonstrator of the relative weight in investment choices for capital in the world, I took the total VC investment in start-ups globally in four consecutive quarters in 2016, which comes to approximately US$165 billion, and compared it with the total capitalization value of NYSE, which was US$18,500 billion in 2016, and accounted for only 27% of global total. Jason Rowley, "Inside the Q2 2017 Global Venture Capital Ecosystem," TechCrunch.com, July 11, 2011, https://techcrunch.com/2017/07/11/inside-the-q2-2017-global-venture-capital-ecosystem.

26. Ian Goldin and Mike Mariathasan, *The Butterfly Effect: How Globalization Creates Systemic Risks, and What to Do about Them* (Princeton, NJ: Princeton University Press, 2015).

27. J. P. Morgan, "Fallout from COVID-19: Global Recession, Zero Interest Rates and Emergency Policy Actions," March 27, 2020, http://www.jpmorgan.com/global/research/fallout-from-covid19.

28. US Government Accountability Office report dated April 20, 2015. Source: http://www.gao.gov/assets/670/669899.pdf.

29. Angela Monaghan, "One in Four UK Families Have Less Than £95 in Savings, Report Finds," *The Guardian*, February 20, 2017, https://www.theguardian.com/society/2017/feb/20/one-in-four-uk-families-have-less-than-95-in-savings-report-finds.

30. Martin Gilens and Benjamin I. Page, "Testing Theories of American Politics: Elites, Interest Groups, and Average Citizens," *Perspectives on Politics* 12, no. 3 (September 2014): 564–581.

31. See https://www.accenture.com/gb-en/insight-artificial-intelligence-future-growth.

32. PricewaterhouseCoopers, "The Macroeconomic Impact of Artificial Intelligence," February 2018, https://www.pwc.co.uk/economic-services/assets/macroeconomic-impact-of-ai-technical-report-feb-18.pdf.

CHAPTER 2

1. George Zarkadakis, *In Our Own Image: The History and Future of Artificial Intelligence* (New York: Pegasus, 2016).

2. Richard Waters, "Microsoft Invests $1bn in OpenAI Effort to Replicate Human Brain," *Financial Times*, July 22, 2019.

3. Zarkadakis, *In Our Own Image*.

4. Steve B. Furber, Francesco Galluppi, Steve Temple, and Luis A. Plana, "The SpiNNaker Project," *Proceedings of the IEEE* 102, no. 5 (2014): 652–655.

5. Jing Pei, Lei Deng, Sen Song, Mingguo Zhao, Youhui Zhang, Shuang Wu, Guanrui Wang, et al. "Towards Artificial General Intelligence with Hybrid Tianjic Chip Architecture," *Nature* 572 (2019): 106–111.

6. Karl Friston, "Learning and Inference in the Brain," *Neural Networks* 16, no. 9 (2003): 1325–1352.

7. Karl Friston, "The Free-Energy Principle: A Rough Guide to the Brain?," *Trends in Cognitive Sciences* 13, no. 7 (2009): 293–301.

8. For example, AlphaGo came up with hitherto unthought of strategies to beat Lee Sedol, a 9th-dan Go master, in Seoul in March 2016.

9. The "black box" problem of deep learning: in an ironic twist of Polanyi's paradox our best AI machines do not "know" how they know what they know, and they cannot explain it to us either.

10. George Zarkadakis, Ravin Jesuthasan, and Tracey Malcolm, "The 3 Ways Work Can Be Automated," *Harvard Business Review*, October 13, 2016.

11. Sonali Basak and Christopher Palmeri, "A Goldman Trading Desk That Once Had 500 People Is Down to Three," Bloomberg, April 30, 2018, https://www.bloomberg.com/news/articles/2018-04-30/goldman-trading-desk-that-once-had-500-people-is-down-to-three.

12. Current AI technology automates many translation tasks (e.g., text-to-text, text-to-voice, voice-to-text). However, as technology advances evermore, translation, as

well as interpretation, tasks will become automated, including efficient and accurate, real-time, voice-to-voice interpretation.

13. Accenture's research estimates a 40% productivity improvement due to AI.

14. Erik Brynjolfsson and Andrew McAfee, *The Second Machine Age: Work, Progress, and Prosperity in a Time of Brilliant Technologies* (New York: W. W. Norton, 2014).

CHAPTER 3

1. Assuming our species is around 200,000 years old, and that the agricultural revolution started roughly 10,000 years ago, humans have been "working" for less than 5% of our evolutionary history.

2. Timothy A. Kohler, Michael E. Smith, Amy Bogaard, Gary M. Feinman, Christian E. Peterson, Alleen Betzenhauser, Matthew Pailes, et al., "Greater Post-Neolithic Wealth Disparities in Eurasia Than in North America and Mesoamerica," *Nature* 551 (November 2017): 619–622.

3. Jared Diamond, *Guns, Germs and Steel: The Fates of Human Societies* (New York: W. W. Norton, 1997).

4. Tellingly, the root of the word "capital" (cap-) means "head," as in the head of an animal. The same root is perhaps more obvious in the word "de-cap-itation."

5. Genesis 3:17–19.

6. Timothy B. Gage and Sharon DeWitte, "What Do We Know about the Agricultural Demographic Transition?" *Current Anthopology* 50, no. 5 (October 2009): 649–655.

7. Richard B. Lee, *The Dobe Ju/'hoansi* (Belmont, CA: Wadsworth, 1984).

8. Karl Zimmerer, "The World Is Stuck on Eating the Same Few Crops, and That's Really Bad for Us All," ScienceAlert, November 30, 2017, https://www.sciencealert.com/fewer-crops-agrobiodiversity-diet-farming-rice-wheat-maze-health-crisis?perpetual=yes&limitstart=1.

9. Marshall Sahlins, *The Original Affluent Society*, June 21, 2002, http://appropriate-economics/materials/Sahlins.pdf.

10. Andrea Komlosy, *Work: The Last 1,000 Years* (London: Verso, 2018).

11. It is important to highlight the difference between "freedom" and "liberty." Adopting a Jeffersonian perspective, freedom usually means to be free *from* something (in this case "work"), whereas liberty usually means to be free *to do* something, for instance to exercise the right to own property, speak one's opinion, choose how to live one's life, choose whom to marry, and so forth. See also Isaiah Berlin, "Two Concepts of Liberty," *Four Essays on Liberty* (Oxford: Oxford University Press, 1969), 118–172.

12. Graph based on data from the Maddison project database: https://www.rug.nl/ggdc/historicaldevelopment/maddison/releases/maddison-project-database-2018. License and funding: Maddison Project Database, version 2018, by Jutta Bolt, Robert Inklaar, Herman de Jong, and Jan Luiten van Zanden is licensed under a Creative Commons Attribution 4.0 International License.

13. "The Labour Share in G20 Economies," report by OECD prepared for the G20 Employment Working Group, Antalya, Turkey (February 26–27, 2015).

14. Ronald Coase, "The Nature of the Firm," *Economica* 4, no. 16 (1937): 386–405.

15. Bourree Lam, "Where Do Firms Go When They Die?," *The Atlantic*, April 12, 2015, https://www.theatlantic.com/business/archive/2015/04/where-do-firms-go-when-they-die/390249.

16. Geoffrey G. Parker, Marshall W. Van Alstyne, and Sangeet Paul Choudary, *Platform Revolution: How Networked Markets Are Transforming the Economy and How to Make Them Work for You* (New York: W. W. Norton, 2016).

17. As, for example, Disney is doing in its theme parks.

18. George Zarkadakis, "Do Platforms Work?," *Aeon*, May 28, 2018.

19. Martin Konrad, "Freelancers Make Up 34 Percent of the U.S. Workforce: Here's How to Find, Hire and Manage Them," *Entrepreneur*, May 24, 2016, https://www.entrepreneur.com/article/275362.

20. Mehreen Khan, "EU Seeks Greater Protection for Gig Economy Workers," *Financial Times*, March 13, 2018, https://www.ft.com/content/dff6d21a-26cc-11e8-b27e-cc62a39d57a0.

21. James Manyika, Susan Lund, Jacques Bughin, Kelsey Robinson, Jan Mischke, and Deepa Mahajan, "Independent Work: Choice, Necessity, and the Gig Economy," McKinsey Global Institute, October 2016, https://www.mckinsey.com/global-themes/employment-and-growth/independent-work-choice-necessity-and-the-gig-economy.

22. This vacuum in labor law is being debated and starting to be addressed in various jurisdictions. An example is the Supreme Court ruling in California in April 2018 that raised the likelihood of companies such as Uber and Lyft categorizing gig workers as employees.

23. Zhang Ruimin and Paul Michelman, "Leading to Become Obsolete," *MIT Sloan Management Review*, June 19, 2017, https://sloanreview.mit.edu/article/leading-to-become-obsolete.

24. Brian J. Robertson, *Holacracy: The Revolutionary Management System That Abolishes Hierarchy* (New York: Penguin, 2014).

25. Sarah Kessler, "How Uber Manages Drivers without Technically Managing Drivers," Fast Company, August 9, 2016, https://www.fastcompany.com/3062622/how-ubers-app-manages-drivers-without-technically-managing-drivers.

26. Carl Benedikt Frey and Michael Osborne, "The Future of Employment: How Susceptible Are Jobs to Computerization?," 2013, https://www.oxfordmartin.ox.ac.uk/publications/the-future-of-employment.

27. The Word Bank, *World Development Report 2016: Digital Dividends,* http://world-bank.org/en/publication/wdr2016.

28. Daron Acemoglu and Pascual Restrepo, "Robots and Jobs: Evidence from US Labor Markets," NBER Working Paper 23285, March 2017.

29. Karen Harris, Austin Kimson, and Andrew Schwedel, "Labor 2030: The Collision of Demographics, Automation, and Inequality," Bain & Co., February 7, 2018, https://www.bain.com/insights/labor-2030-the-collision-of-demographics-automation-and-inequality.

30. Rodney Brooks, "The Productivity Gain: Where Is It Coming from and Where Is It Going To?," February 25, 2018, https://rodneybrooks.com/the-productivity-gain-where-is-it-coming-from-and-where-is-it-going-to.

31. Ravin Jesuthasan, Tracey Malcolm, and George Zarkadakis, "Automation Will Make Us Rethink What a 'Job' Really Is," *Harvard Business Review*, October 12, 2016.

32. Richard Wike and Bruce Stokes, "In Advanced and Emerging Economies Alike, Worries about Job Automation," Pew Research Center, September 13, 2018, http://www.pewglobal.org/2018/09/13/in-advanced-and-emerging-economies-alike-worries-about-job-automation.

33. Axios labor data research 2018, https://www.axios.com/most-jobs-created-since-recciu-1536269032-13ccc866-5fb0-44e8-bd14-286ae09c296f.html.

34. Data from https://fred.stlouisfed.org/series/MANEMP.

CHAPTER 4

1. Jon Agar, *The Government Machine: A Revolutionary History of the Computer* (Cambridge, MA: MIT Press, 2003).

2. Agar, *The Government Machine.*

3. Virginia Eubanks, *Automating Inequality: How High-Tech Tools Profile, Police, and Punish the Poor* (New York: St. Martin's Press, 2018).

4. Eubanks, *Automating Inequality.*

5. Jeff Larson, Surya Mattu, Lauren Kirchner, and Julia Angwin, "How We Analyzed the COMPAS Recidivism Algorithm," ProPublica, May 23, 2016, https://www.propublica.org/article/how-we-analyzed-the-compas-recidivism-algorithm.

6. Kristian Lum and William Isaac, "To Predict and Serve?," *Significance* 13, no. 5 (October 7, 2016): 14–19, https://rss.onlinelibrary.wiley.com/doi/full/10.1111/j.1740-9713.2016.00960.x.

7. Safiya Umoja Noble, *Algorithms of Oppression: How Search Engines Reinforce Racism* (New York: NYU Press, 2018).

8. Elisa Shearer and Katerina Eva Matsa, "New Use across Social Media Platforms 2018," Pew Research Center, September 10, 2018, https://www.journalism.org/2018/09/10/news-use-across-social-media-platforms-2018.

9. Carole Cadwalladr and Emma Graham-Harrison, "Revealed: 50 Million Facebook Profiles Harvested for Cambridge Analytica in Major Data Breach," *The Guardian*, March 17, 2018, accessed via:Raja, https://www.theguardian.com/news/series/cambridge-analytica-files.

10. Raja Chatila, *Ethically Aligned Design*, IEEE Standards Association, January 21, 2020, https://standards.ieee.org/develop/indconn/ec/autonomous_systems.html.

11. Manuel Velasquez, Claire Andre, Thomas Shanks, and Michael J. Meyer, "The Common Good," *Issues in Ethics* 5, no. 1 (Spring 1992).

12. https://www.wyden.senate.gov/imo/media/doc/Algorithmic%20Accountability%20Act%20of%202019%20Bill%20Text.pdf.

13. Big tech companies are exerting enormous influence in shaping the landscape of AI ethics and regulation. In 2019 the US National Science Foundation launched a US$7.6 million funding program on Fairness in Artificial Intelligence, funded by Amazon, as reported in Yochai Benkler, "Don't Let Industry Write the Rules of AI," *Nature* 569, no. 7755 (2019): 161. In the same year Facebook invested US$7.5 million in a center of ethics and AI at the Technical University of Munich, Germany.

14. Ludwig von Mises, *Economic Calculation in the Socialist Commonwealth* (Auburn, AL: Mises Institute, 1990).

15. Geoffrey M. Hodgson, "Socialism against Markets? A Critique of Two Recent Proposals," *Economy and Society* 27, no. 4 (November 1998): 407–433.

16. Eden Medina, *Cybernetic Revolutionaries: Technology and Politics in Allende's Chile* (Cambridge, MA: MIT Press, 2011).

17. Feng Xiang, "AI Will Spell the End of Capitalism," *The Washington Post*, May 3, 2018, https://www.washingtonpost.com/news/theworldpost/wp/2018/05/03/end-of-capitalism/?noredirect=on&utm_term=.ea05123dfc43.

18. Binbin Wang and Xiaoyan Li, "Big Data, Platform Economy, and Market Competition: A Preliminary Construction of Plan-Oriented Market Economy System in the Information Era," *World Review of Political Economy* 8, no. 2 (Summer 2017): 138–161.

19. See the BBC report on Chinese reeducation camps in Xinjiang. John Sudworth, "China Blog: Searching for Truth in China's Uighur 'Re-education' Camps," *BBC News*, June 21, 2019, https://www.bbc.com/news/blogs-china-blog-48700786.

20. Douglas Ernst, "China Touts 'Social Credit' System to Deny Travel: 'Once Untrustworthy, Always Restricted,'" *The Washington Times*, March 16, 2018, https://www.washingtontimes.com/news/2018/mar/16/china-touts-social-credit-system-to-deny-travel-on.

21. Isaac Asimov, "Franchise," *If: Worlds of Science Fiction*, August 1955.

22. Christina Larson, "Who Needs Democracy When You Have Data?," *MIT Technology Review*, August 20, 2018, https://www.technologyreview.com/s/611815/who-needs-democracy-when-you-have-data.

23. "All Watched Over by Machines of Loving Grace" was the title of a poem by Richard Brautigan (1935–1984).

24. Darryl Campbell, "The Many Human Errors That Brought Down the Boeing 737 Max," *The Verge*, May 1, 2019.

25. Edward Geist and Andrew J. Lohn, "How Might Artificial Intelligence Affect the Risk of Nuclear War?," Rand International, Security 2040 Project, 2018, https://www.rand.org/pubs/perspectives/PE296.html.

26. Edward Geist and Andrew J. Lohn, "How Might Artificial Intelligence Affect the Risk of Nuclear War?," Perspectives, 2018, https://www.rand.org/pubs/perspectives/PE296.html.

CHAPTER 5

1. "Websites Blocked in Mainland China," Wikipedia, https://en.wikipedia.org/wiki/Websites_blocked_in_mainland_China.

2. Nicholas Wright, "How Artificial Intelligence Will Reshape the Global Order," *Foreign Affairs*, July 10, 2018, https://www.foreignaffairs.com/articles/world/2018-07-10/how-artificial-intelligence-will-reshape-global-order.

3. Li Tao, "Malaysian Police Wear Chinese Start-Up's AI Camera to Identify Suspected Criminals," *South China Morning Post*, April 20, 2018, https://www.scmp.com/tech/social-gadgets/article/2142497/malaysian-police-wear-chinese-start-ups-ai-camera-identify.

4. Louise Matsakis, "What Happens if Russia Cuts Itself Off from the Internet," *Wired*, February 12, 2019.

5. Various primary sources as listed in "Internet Censorship in Russia" article, Wikipedia, http://en.wikipedia.org/wiki/internet_censorship_in_Russia.

6. Shannon Van Sant, "Russia Criminalizes the Spread of Online News That 'Disrespects' the Government," NPR, March 18, 2019, https://www.npr.org/2019/03/18/704600310/russia-criminalizes-the-spread-of-online-news-which-disrespects-the-government?t=1564467761076.

7. Wright, "How Artificial Intelligence Will Reshape the Global Order."

8. B. F. Skinner, *Beyond Freedom and Dignity* (New York: Knopf, 1971).

9. Shoshana Zuboff, *The Age of Surveillance Capitalism: The Fight for a Human Future at the New Frontier of Power* (New York: Public Affairs, 2019).

10. Mostly through acquisitions—for example, Facebook acquiring WhatsApp.

11. See, for example, how Facebook is becoming a bank with the launch of Libra, or how Google is disrupting the automobile industry with Waymo.

12. Daniel Castro, "The EU's Right to Be Forgotten Is Now Being Used to Protect Murderers," Centre for Data Innovation, September 21, 2018, https://www.datainnovation.org/2018/09/the-eus-right-to-be-forgotten-is-now-being-used-to-protect-murderers.

13. Cecilia Kang, "FTC Approves Facebook Fine of about $5 Billion," *The New York Times*, July 12, 2019, https://www.nytimes.com/2019/07/12/technology/facebook-ftc-fine.html.

14. Charles Riley and Ivana Kottasová, "Europe Hits Google with a Third, $1.7 Billion Antitrust Fine," CNN, March 20, 2019, https://edition.cnn.com/2019/03/20/tech/google-eu-antitrust/index.html.

15. Wright, "How Artificial Intelligence Will Reshape the Global Order."

16. Stef W. Kight, "Exclusive Poll: Young Americans Are Embracing Socialism," Axios, March 10, 2019, https://www.axios.com/exclusive-poll-young-americans-embracing-socialism-b051907a-87a8-4f61-9e6e-0db75f7edc4a.html.

17. Virginia Furness, Sujata Rao, and Julien Ponthus, "Corbyn-Proof? British Water, Power Firms Take Nationalisation Precautions," Reuters, April 29, 2019, https://uk.reuters.com/article/uk-britain-eu-labour-privatisation-analy/corbyn-proof-british-water-power-firms-take-nationalisation-precautions-idUKKCN1S50BZ.

18. Jochen Bittner, "Why Is Socialism Coming Back in Germany?," *The New York Times*, May 2, 2019, https://www.nytimes.com/2019/05/02/opinion/germany-socialism.html.

19. Liberal democracies today are "mixed economies" where the state and also international trade agreements and institutions exercise a high degree of control and regulation over so-called "free markets," skewing free market dynamics via subsidies and public investment, while high taxation of personal income and corporate profits is used to redistribute wealth and fund public services, such as education and health care.

20. Friedrich A. Hayek, *The Road to Serfdom* (Chicago: University of Chicago Press, 1944).

21. Hayek, *The Road to Serfdom*.

22. Josh Barro, "Lessons from the Decades Long Upward March of Government Spending," *Forbes*, April 16, 2012, https://www.forbes.com/sites/joshbarro/2012/04/16/lessons-from-the-decades-long-upward-march-of-government-spending/#13b137227201.

23. Calculated on 2019 data. Source: https://tradingeconomics.com/european-union/government-spending.

24. Steve Robinson, "Growth of Government Assistance Adds to National Debt," Maine Wire, January 13, 2103, https://www.themainewire.com/2013/01/growth-government-assistance-adds-national-debt.

25. Rutger Bregman, *Utopia for Realists: And How We Can Get There* (London: Bloomsbury, 2018).

26. Evelyn L. Forget, "The Town with No Poverty: The Health Effects on a Canadian Guaranteed Annual Income Field Experiment," *Canadian Public Policy*, 37, no. 3 (2011): 283–305.

27. Carrie Arnold, "Money for Nothing: The Truth about Universal Basic Income," *Nature*, News Feature, May 30, 2018, https://www.nature.com/articles/d41586-018-05259-x.

28. Clem Chambers, "Money and Markets: Universal Basic Income—Not So Revolutionary," theiet.org, April 16, 2019, https://eandt.theiet.org/content/articles/2019/04/money-markets-universal-basic-income.

29. Chambers, "Money and Markets."

30. Chambers, "Money and Markets."

31. Support for government intervention to compensate for income loss is widespread among US tech leaders. For example, Bill Gates is in favor of a "robot tax." In 2018 Y Combinator announced a long-term study in UBI.

32. Zuboff, *The Age of Surveillance Capitalism.*

33. Bryan Caplan, *The Myth of the Rational Voter: Why Democracies Choose Bad Policies* (Princeton, NJ: Princeton University Press, 2007).

34. Daniel Kahneman, *Thinking Fast and Slow* (New York: Farrar, Straus and Giroux, 2011).

35. Tim Harford, "Referendums Break Democracies So Best to Avoid Them," *Financial Times,* March 2, 2018.

36. Hans-Hermann Hoppe, *Democracy: The God That Failed* (Livingston, NJ: Transaction, 2001).

37. The same research by Pew has also found a 49%–46% split in favor of being ruled by experts, while 26% preferred to be rules by a "strong leader" and 24% by the military. Pew Research Center, "Widespread Support for Representative and Direct Democracy, but Many Are Also Open to Nondemocratic Alternatives," October 12, 2017, http://www.pewglobal.org/2017/10/16/globally-broad-support-for-representative-and-direct-democracy/pg_2017-10-16_global-democracy_0-01.

CHAPTER 6

1. The Meeting of Minds project was funded by the European Commission and the King Baudouin Foundation and took place between 2005 and 2006.

2. Terms such as "epistocracy" or "noocracy" are used to describe rule by experts, which broadly corresponds to what ancient Greeks meant by rule by the best (the "aristoi"), hence the word "aristocracy."

3. The nine countries that participated in the Meeting of Minds project were Belgium, Denmark, France, Germany, Greece, Hungary, Ireland, the Netherlands, and the United Kingdom. In total 126 EU citizens took part.

4. This restriction included professionals such as neuroscientists, neurobiologists, neurologists, psychiatrists, and psychologists.

5. The citizens who were selected would be interacting with a wide spectrum of stakeholders, including patient groups and other special interest groups, on the basis of suggestions but also as they saw fit.

6. The six themes used for deliberation and recommendations were Regulation and Control, Normalcy versus Diversity, Public Information, Pressure from Economic Interests, Equal Access to Treatment, and Freedom of Choice.

7. The Greek word for someone who thinks, acts and cares only for himself or herself is ιδιώτης (*idiotes*), which the English word "idiot" is derived from.

8. This crucial observation on the sine qua non for a citizen democracy is also made in Josiah Ober, *Demopolis: Democracy before Liberalism in Theory and Practice* (Cambridge: Cambridge University Press, 2017).

9. Joseph M. Bessette, "Deliberative Democracy: The Majority Principle in Republican Government," in *How Democratic Is the Constitution?,* ed. Robert A. Goldwin (Washington, DC: AEI Press, 1980), 102–116.

10. The Athenian Citizen Assembly was called "Ecclesia of the Demos." "Ecclesia" (*Εκκλησία*) in Greek means the gathering.

11. Similarly to the selection for jury duty.

12. Remesh's typical use case is to deliver cost-efficient "focus groups" on the web for commercial purposes. However, their application can also be used to facilitate deliberations in a citizen assembly. See https://remesh.ai.

13. Claudia Chwalisz, *The People's Verdict: Adding Informed Citizen Voices to Public Decision-Making* (Lanham, MD: Rowman & Littlefield, 2017).

14. The Irish Referendum took place on May 26, 2018. Citizens were asked to decide whether parliament should repeal the 1983 eighth amendment that prohibited abortion except in exceptional circumstances. Nearly two-thirds of Irish voters opted for a repeal.

15. You can watch a short film of the citizen assembly deliberations here: https://www.allhandsondoc.com.

16. Chwalisz, *The People's Verdict.*

17. Ji-Bum Chung, "Let Democracy Rule Nuclear Energy," *Nature* 555, no. 415 (March, 22, 2018).

18. National referenda to maintain or shut down nuclear energy have been held in Sweden (1980), Italy (1987), and Switzerland (1990).

19. The *deliberative poll* is a technique developed by James Fishkin, professor of communication at Stanford University. This technique was used at the Meeting of Minds project, as described in chapter 5 of *Cyber Republic.*

20. See https://granddebat.fr.

21. Beyond citizen assemblies, the Grand Débat included town hall meetings, complaint books, mobile desks in train stations and post offices, online suggestion forms, and four stakeholder conferences in Paris.

22. Renaud Thillaye, "Is Macron's Grand Débat a Democratic Dawn for France?," Carnegie Europe, April 29, 2019, https://carnegieeurope.eu/2019/04/26/is-macron-s-grand-d-bat-democratic-dawn-for-france-pub-79010.

23. Thillaye, "Is Macron's Grand Débat a Democratic Dawn for France?"

24. Claudia Chwalisz and David Reybrouck, "Macron's Sham Democracy," Politico, September 9, 2018, https://www.politico.eu/article/macron-populism-sham-democracy-plans-to-revamp-decision-making-disappointing.

25. Ober, *Demopolis.*

26. Ober, *Demopolis.*

27. This is what happened with genetically modified organisms (GMOs) in Europe. Decisions on funding GMOs were made among experts and politicians behind closed doors in elite deliberations until the citizens found out through the reporting, and distorting lens, of various media outlets. The popular reaction that ensued led to the hasty folding of GMO research in Europe and the banning of GMOs. This result damaged the experts, but it also damaged the citizens by keeping the costs of agricultural produce high and jeopardizing the resilience of food stocks in the face of climate change, desertification, and crop diseases.

CHAPTER 7

1. Anton Korinek, "Integrating Ethical Values and Economic Value to Steer Progress in Artificial Intelligence," NBER Working Paper Series, Working Paper 26130 (2019), http://www.nber.org/papers/w26130.

2. Daron Acemoglu and Pascual Restrepo, "The Wrong Kind of AI? Artificial Intelligence and the Future of Labor Demand," NBER Working Paper No. 25682, March 2019, https://www.nber.org/papers/w25682.

3. William James Fox, "Human-Level Artificial Intelligence Could Be Achieved 'within Five to Ten Years,' Say Experts," futuretimeline, September 25, 2018, https://www.futuretimeline.net/blog/2018/09/25.htm.

4. Nick Bostrom, *Superintelligence: Paths, Dangers, Strategies* (Oxford: Oxford University Press, 2014).

5. Bostrom, *Superintelligence.*

6. Cybernetics has given birth to several contemporary scientific disciplines, including systems science and the science of complex adaptive systems, which study the behavior of complex systems such as physical, computational, biological, and social systems.

7. Reto Bisaz, Alessio Travaglia, and Cristina Alberini, "The Neurobiological Bases of Memory Formation: From Physiological Conditions to Psychopathology," *Psychopathology* 46, no. 6 (2014): 347–356.

8. Bisaz, Travaglia, and Alberini, "The Neurobiological Bases of Memory Formation."

9. Norbert Wiener, *The Human Use of Human Beings* (Boston: Houghton Muffin, 1950).

10. John McCarthy, Marvin L. Minsky, Nathaniel Rochester, and Claude E. Shannon, "A Proposal for the Dartmouth Summer Research Project on Artificial Intelligence," August 31, 1955, http://jmc.stanford.edu/articles/dartmouth/dartmouth.pdf.

11. George W. Ernst and Allen Newell, *GPS: A Case Study in Generality and Problem Solving* (New York: Academic Press, 1969).

12. David Silver, Thomas Hubert, Julian Schrittwieser, Ioannis Antonoglou, Matthew Lai, Arthur Guez, Marc Lanctot, et al., "Mastering Chess and Shogi by Self-Play with a General Reinforcement Learning Algorithm," 2017, arxiv.org/abs/1712.01815.

13. As discussed in chapter 2 ("Machines That Think"), one of the most well-funded approaches to AGI today is envisaging using deep neural networks but with massive computing power.

14. Silver et al., "Mastering Chess."

15. Vivi Nastase, Rada Mihalcea, and Dragomir R. Radev, "A Survey of Graphs in Natural Language Processing," *Natural Language Engineering* 21, no. 5 (2015): 665–698.

16. Gordon Pask, *Conversation Theory: Applications in Education and Epistemology* (Amsterdam: Elsevier, 1976).

17. Paul Pangaro and Hugh Dubberly, "What Is Conversation? How Can We Design for Effective Conversation?," 2009, http://www.dubberly.com/articles/what-is-conversation.html.

18. OpenAI, "Better Language Models and Their Implications," February 14, 2019, https://openai.com/blog/better-language-models.

19. Nastase, Mihalcea, and Radev, "A Survey of Graphs."

20. George Zarkadakis, "How Next Generation Search Will Make the Web More Equal," World Economic Forum Agenda, October 14, 2015, https://www.weforum.org/agenda/2015/10/how-next-generation-search-will-make-the-web-more-equal.

21. Autopoesis means "self-creation" in Greek.

22. Peter Harries-Jones, "The Self-Organizing Polity: An Epistemological Analysis of Political Life by Laurent Dobuzinskis," *Canadian Journal of Political Science* 21, no. 2 (June 1988): 431–433.

CHAPTER 8

1. Nakamoto vanished from the public sphere in 2009, and his identity remains unknown to this date. Perhaps that was the best move that he (if indeed Nakamoto is a "he") could have made, given the global media's attention on bitcoin's notoriety, and his personal wealth of bitcoins estimated to be worth around US$16 billion (in January 2018 prices), making him one of the richest people in the world.

2. For a brief overview of the cypherpunk movement, see James Bridle's introduction to *The White Paper* by Satoshi Nakamoto, ed. Jaya Klara Brekke and Ben Vickers (London: Ignota, 2019).

3. One of the most successful implementations of blockchain has been a marketplace for digital images of cats called "cryptokitties" (see cryptokitties.co), which goes to show the limitless potential of creating wealth out of one's imagination.

4. Leslie Lamport, Robert Shostak, and Marshall Pease, "The Byzantine Generals Problem," *ACM Transactions on Programming Languages and Systems* 4, no. 3 (July 1982): 382–401.

5. Leslie Lamport has given an explanation for choosing the term "Byzantine" in Lamport, Shostak, and Pease, "The Byzantine Generals Problem."

6. In June 2018 block miners were awarded 12.5 bitcoins. This rate will decrease over time (see also note 8 below).

7. There are only 21 million bitcoins to be mined in total, and the difficulty of mining them increases over time.

8. Anami Nguyen, "Intro to Cryptoeconomics—Part I," Medium.com, June 11, 2018, https://medium.com/@anaminguyen/intro-to-cryptoeconomics-part-1-b2527775bc9c.

9. Private cryptonetworks are also called "permissioned" as opposed to public cryptonetworks that are called "permissionless."

10. See European Council's position on digital taxation here: https://www.consilium.europa.eu/en/policies/digital-taxation.

11. CAC: how much money needs to be invested in order to convert someone into a loyal user of a platform.

12. ICO: a way to raise money for a business by issuing digital securities ("utility tokens") to investors. ICOs were used by many, often dubious, start-ups who managed to raise considerable amounts of (fiat) cash solely on the basis of a "vision" written in a white paper and without having built anything at all. For example, in May 2018 a Cayman Islands–based start-up called Block.one, offering a cryptocurrency called "eos," raised US$4 billion without any product.

13. The Gartner Hype Cycle distinguishes four stages in the evolution of a technology: trigger (when the technology first appears); hype (when the technology is hyped in the news, usually causing high expectations); trough of disillusionment (following hype, as the technology is not mature enough to deliver on the promised expectations); and plateau of productivity (when the technology actually delivers real value but does not enjoy as much media attention anymore).

14. Satoshi Nakamoto, "Bitcoin: A Peer-to-Peer Electronic Cash System," bitcoin.org, 2008, https://bitcoin.org/bitcoin.pdf.

15. Costs for mining bitcoins vary from US$531 to US$26,170, depending on the country. The United States ranks as the 40th cheapest to mine a single bitcoin, with an average cost of US$4,758 (in 2018).

16. Belarus announced in May 2019 that it will begin mining bitcoins by leveraging surplus from its nuclear energy plants.

17. As reported in various Western media, including *Fortune* magazine; see Naomi Xu Elegant, "Why China's Digital Currency Is a 'Wake-Up Call' for the US," *Fortune*, November 1, 2019, https://fortune.com/2019/11/01/china-digital-currency-libra-wakeup-call-us.

18. As reported in various media, including https://bitcoinist.com/bank-of-england-mark-carney-suggests-global-cryptocurrency.

19. Jacques Bughin, Jeongmin Seong, James Manyika, Michael Chui, and Raoul Joshi, "Modeling the Impact of AI on the World Economy," McKinsey Report, September 2018, https://www.mckinsey.com/featured-insights/artificial-intelligence/notes-from-the-frontier-modeling-the-impact-of-ai-on-the-world-economy.

20. Bughin et al., "Modeling the Impact of AI."

CHAPTER 9

1. Richard Henderson and Patrick Temple-West, "Group of Corporate Leaders Ditches Shareholder-First Mantra," *Financial Times*, August 19, 2019, https://www.ft.com/content/e21a9fac-c1f5-11e9-a8e9-296ca66511c9.

2. Edelman, "In Brands We Trust?" Trust Barometer special report, 2019, https://www.edelman.com/sites/g/files/aatuss191/files/2019-06/2019_edelman_trust_barometer_special_report_in_brands_we_trust.pdf.

3. Mark Freeman, Robin Pearson, and James Taylor, *Shareholder Democracies? Corporate Governance in Britain and Ireland before 1850* (Chicago: University of Chicago, 2012).

4. Rob McQueen, *A Social History of Company Law: Great Britain and the Australian Colonies*, (Farnham, UK: Ashgate Publishing, 2009).

5. McQueen, *A Social History of Company Law*.

6. Paul Johnson, *Making the Market: Victorian Origins of Corporate Capitalism* (Cambridge: Cambridge University Press, 2010).

7. Colin Mayer, *Prosperity: Better Business Makes the Greater Good* (Oxford: Oxford University Press, 2018).

8. Data quoted in https://en.wikipedia.org/wiki/List_of_public_corporations_by_market_capitalization.

9. I am using the terms "web 3.0 platforms," "cryptoplatforms," and "cryptonetworks" interchangeably throughout the book.

10. George Zarkadakis, "Do Platforms Work?" Aeon, 2018, https://aeon.co/essays/workers-of-the-world-unite-on-distributed-digital-platforms.

11. For a discussion on forking and cryptogovernance, see Bruno Rodrigues, "In Crypto Economy, Governance Is Key!," Medium, December 28, 2017, https://medium.com/insightsaltaperformance/in-crypto-economy-governance-is-key-8fb5430f7972.

12. In this example I assume dependency on fossil fuels to make the point of interconnectedness; naturally, one can replace "fossil fuel" with "electricity."

13. See the blog post on cryptogovernance by Steven McKie, "The Crypto Governance Manifesto," https://medium.com/blockchannel/the-crypto-governance-manifesto-2326e72dc3d0.

14. Ellie Rennie, "The Radical DAO Experiment," *Winburne News*, May 12, 2016.

15. "SEC Issues Investigative Report Concluding DAO Tokens, a Digital Asset, Were Securities," Press Release, July 25, 2017, accessed via http://sec.gov/news/press-release/2017-131.

16. See http://lazooz.org.

17. Jacquelyn Cheok, "Blockchain-Based Ride-Hailing App TADA Makes Singapore Debut," *Business Times*, July 27, 2018, https://www.businesstimes.com.sg/startups/blockchain-based-ride-hailing-app-tada-makes-singapore-debut.

18. Zheping Huang, "The Ride-Hailing Pioneer Thinks Blockchain Can Solve the Ride-Hailing Safety Crisis in China," *South China Morning Post*, August 30, 2018, https://www.scmp.com/tech/enterprises/article/2161900/chinese-ride-hailing-pioneer-returns-blockchain-app-boost-safety.

19. See https://www.suppapp.com.

20. Rishi Iyengar, "Migrant Workers Will Send Home $450 Billion This Year," CNN, June 15, 2017, http://money.cnn.com/2017/06/15/news/economy/migrant-workers-global-remittances/index.html.

21. Email exchange between the author and Jordan Murray, dated July 31, 2018.

22. Charles I. Jones and Christopher Tonetti, "Nonrivalry and the Economics of Data," 2018, http://christophertonetti.com/files/papers/JonesTonetti_DataNonrivalry.pdf.

23. A subset of "personal data" is also often referred to as "personally identifiable information" (PII) and is subject to data regulation. PII data would include data such as social security numbers, bank account, passport number, email, and so forth.

24. Nidhi Kalra and Susan M. Paddock, "Driving to Safety: How Many Miles of Driving Would It Take to Demonstrate Autonomous Vehicle Reliability," RAND Corporation research report, 2018, https://www.rand.org/content/dam/rand/pubs/research_reports/RR1400/RR1478/RAND_RR1478.pdf.

25. Such as PII data.

26. See theodi.org.

27. See https://oceanprotocol.com.

28. Market data in 2019 quoted in Jeffrey Burt, "AWS Still King in Public Cloud, While Azure Grows Fastest, IBM Falls," Channel Partners, n.d., https://www.channelpartnersonline.com/2019/02/07/azure-still-king-in-public-cloud-while-azure-grows-fastest-ibm-falls.

29. PricewaterhouseCoopers, "PwC's Global Artificial Intelligence Study: Exploiting the AI Revolution," 2017, https://www.pwc.com/gx/en/issues/data-and-analytics/publications/artificial-intelligence-study.html.

30. See https://dadi.cloud/en.

31. "Introduction to Distributed Generation" Virginia Tech, 2007, accessed via https://www.dg.history.vt.edu/ch1/introduction.html.

32. Kate Cummins, "The Rise of Additive Manufacturing," *The Engineer*, May 23, 2010. http://theengineer.co.uk/the-rise-of-additive-manufacturing.

33. Robert McDougall, Paul Kristiansen, and Romina Rader, "Small-Scale Urban Agriculture Results in High Yields but Requires Judicious Management of Inputs to Achieve Sustainability, *Proceedings of the National Academy of Sciences of the United States of America* 116, no. 1 (January 2, 2019): 129–134.

34. George Zarkadakis, "The Economy Is More a Messy Fractal Living Thing, Than a Machine," *Aeon Magazine*, October 13, 2016.

CHAPTER 10

1. World Bank, https://www.worldbank.org/en/topic/urbandevelopment/overview.

2. United Nations, "68% of the World Population Projected to Live in Urban Areas by 2050, Says UN," May 16, 2018, https://www.un.org/development/desa/en/news/population/2018-revision-of-world-urbanization-prospects.html.

3. Luis Bettencourt and Geoffrey West, "A Unified Theory of Urban Living," *Nature* 467 (October 20, 2010): 912–913.

4. The definition of *civitas* comes from Cicero (Somn. Scip. c3), and it means city in the sense of the polity of citizens united under a common law; it thus corresponds directly to the Greek word "polis."

5. Richard Dobbs, Sven Smit, Jaana Remes, James Manyika, Charles Roxburgh, and Alejandra Restrepo, "Urban World: Mapping the Economic Power of Cities," McKinsey, March 2011, https://www.mckinsey.com/featured-insights/urbanization/urban-world-mapping-the-economic-power-of-cities.

6. Bettencourt and West, "A Unified Theory of Urban Living."

7. See https://www.sidewalklabs.com.

8. Jane Wakefield, "The Google City That Has Angered Toronto," BBC News, May 18, 2019, https://www.bbc.co.uk/news/technology-47815344.

9. Wakefield, "The Google City That Has Angered Toronto."

10. Bettencourt and West, "A Unified Theory of Urban Living."

11. Alyssa Harvey Dawson, "We Believe Quayside Can Set a New Model for Responsible Use of Data in Cities—Anchored by an Independent Civic Data Trust," Sidewalk Labs, October 15, 2018, https://www.sidewalklabs.com/blog/an-update-on-data-governance-for-sidewalk-toronto.

12. In 2019 the Toronto Region Board of Trade suggested that the Toronto Public Library should take on the role of the data trust.

13. Open Data Institute, "Data Trusts: Lessons from Three Pilots," ODI, April 15, 2019, https://theodi.org/article/odi-data-trusts-report.

14. Paul Copping, Jarmo Eskelinen, Ken Figueredo, Michael Fisher, and Lindsay Frost, "The 4th Industrial Revolution and Municipal CEO," ETSI White Paper No. 26, 2018.

15. See https://www.kiva.org/protocol.

16. As quoted in Neil Munshi and Hannah Murphy, "Sierra Leone's Thumbprint Breakthrough to Sign Up Unbanked," *Financial Times*, August 21, 2019.

17. World Bank, "Inclusive and Trusted Digital ID Can Unlock Opportunities for the World's Most Vulnerable," August 14, 2019, https://www.worldbank.org/en/news/immersive-story/2019/08/14/inclusive-and-trusted-digital-id-can-unlock-opportunities-for-the-worlds-most-vulnerable.

18. Olivia White, Anu Madgavkar, James Manyika, Deepa Mahajan, Jacques Bughin, Mike McCarthy, and Owen Sperling, "Digital Identification: A Key to Inclusive Growth," McKinsey, April 2019, https://www.mckinsey.com/business-functions/mckinsey-digital/our-insights/digital-identification-a-key-to-inclusive-growth.

19. World Bank, "Global ID Coverage, Barriers, and Use by the Numbers: Insights from the ID4D-Findex Survey," n.d., http://documents.worldbank.org/curated/en/953621531854471275/pdf/Global-ID-Coverage-by-the-Numbers-Insights-from-the-ID4D-Findex-Survey.pdf.

20. Kenneth T. Frank, Brian Petrie, Jae S. Choi, and William C. Leggett, "Trophic Cascades in a Formerly Cod-Dominated Ecosystem," *Science* 308, no. 5728 (2005): 1621–1623.

21. At about the same time the Homestead Act in the US privatized lands that belonged to Native Americans.

22. Karl Polanyi, *The Great Transformation* (New York: Farrar & Rinehart, 1994).

23. Polanyi, *The Great Transformation.*

24. Elinor Ostrom, *Governing the Commons: The Evolution of Institutions for Collective Action* (Cambridge: Cambridge University Press, 1990).

25. Ostrom, *Governing the Commons.*

26. Demetris Loizos, "Economic History Problems of 18th c. Ottoman Greece: The Case of Ampelakia in Thessaly," *Anistoriton*, E011 (April 2001).

27. The full list of Ostrom's design principles is as follows:

1. Clearly defined (clear definition of the contents of the common pool resource and effective exclusion of external un-entitled parties);
2. The appropriation and provision of common resources that are adapted to local conditions;
3. Collective-choice arrangements that allow most resource appropriators to participate in the decision-making process;
4. Effective monitoring by monitors who are part of or accountable to the appropriators;
5. A scale of graduated sanctions for resource appropriators who violate community rules;
6. Mechanisms of conflict resolution that are cheap and of easy access;
7. Self-determination of the community recognized by higher-level authorities; and
8. In the case of larger CPR, organization in the form of multiple layers of nested enterprises, with small local CPRs at the base level. (Ostrom, *Governing the Commons*)

28. Guillaume Chapron, "The Environment Needs Cryptogovernance," *Nature* 545 (May 25, 2017): 403–405.

29. Georg Eder, "Digital Transformation: Blockchain and Land Titles," in 2019 OECD Global Anti-Corruption & Integrity Forum, Paris, March 20–21, 2019.

30. Eder, "Digital Transformation."

31. Chapron, "The Environment Needs Cryptogovernance."

EPILOGUE

1. As currently supported by the Bank of England—see the discussion in chapter 8 on SHC.

2. United Nations, "World Population Projected to Reach 9.7 Billion by 2050," July 29, 2015, http://www.un.org/en/development/desa/news/population/2015-report.html.

3. Nicholas Christakis, *Blueprint: The Evolutionary Origins of a Good Society* (New York: Little, Brown Spark, 2019).

4. David C. Lahti and Bret S. Weinstein, "The Better Angels of Our Nature: Group Stability and the Evolution of Moral Tension," *Evolution & Human Behavior* 26, no. 1 (January 2005): 47–63.

5. Patricia S. Churchland, *Conscience: The Origins of Moral Intuition* (New York: W. W. Norton, 2019).

6. Henry Foy, "Why Russian Teenager Olya Misik Is Defying Putin's Rule," *Financial Times*, August 20, 2019, https://eblnews.com/video/why-russian-teenager-olya-misik-defying-putins-rule-731087.

INDEX

Note: The letter *f* following a page number denotes a figure.